# Machine Learning
## Mastery

*Complete Guide for Beginners*

"Artificial Intelligence is the New Electricity"

— Andrew Ng

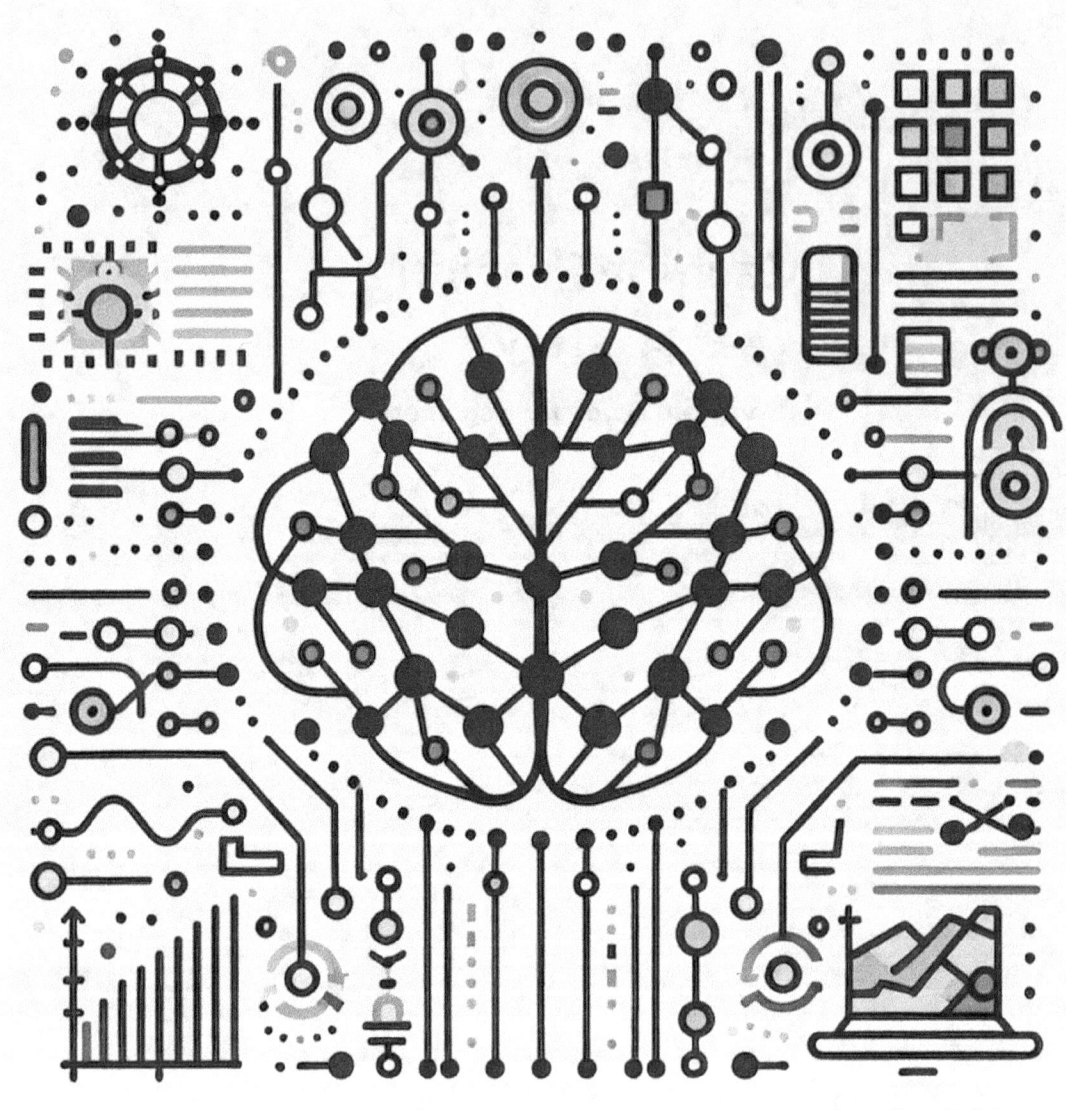

# Table of Contents

**Introduction**    11
    1. What is Machine Learning?
- Definition and basic concepts
- Differences between Machine Learning, Artificial Intelligence and Deep Learning    12
- Applications of Machine Learning in daily life

    2. Why Machine Learning is Important    12
- Impact on industry and society
- Examples of success in various fields

**Chapter 1: Basic Concepts and Terminology**    15
    1. Data and Data Sets
- Type of data
- Data sources
- Data preparation and cleaning

    2. Variables    15
- Dependent and independent variables
- Types of variables (categorical, continuous, ordinal)    16

    3. Types of Learning    16
- Supervised learning
- Unsupervised learning
- Semi-supervised learning
- reinforcement learning

    4. Cost Function and Optimization    16
- Definition of the cost function
- Optimization algorithms (gradient descent, etc.)    17

**Chapter 2: Data Preparation**    21
    1. Data Cleaning
- Handling missing values
- Detection and removal of outliers    22

    2. Data Transformation    23
- Normalization and standardization
- Coding of categorical variables    24

    3. Feature Selection and Extraction    26
- Feature selection techniques
- Feature Extraction Techniques

## Chapter 3: Machine Learning Algorithms — 29
1. Linear and Logistic Regression
   - Fundamentals and mathematics behind regression
   - Practical examples and applications — 30

2. Decision Trees and Random Forests — 34
   - Decision Tree Fundamentals
   - Random forests and how they work — 36

3. Support Vector Machines (SVM)** — 38
   - SVM Fundamentals
   - Applications and practical examples

4. Neural Networks and Deep Learning — 40
   - Fundamentals of neural networks
   - Common architectures (MLP, CNN, RNN)
   - Introduction to deep learning

## Chapter 4: Model Evaluation and Validation — 45
1. Performance Metrics
   - Metrics for regression problems
   - Metrics for classification problems — 46

2. Cross Validation — 48
   - Cross validation techniques
   - Implementation and examples

3. Overfitting y Underfitting — 49
   - Definition and causes
   - Techniques to mitigate them

## Chapter 5: Practical Implementation — 53
1. Popular Tools and Libraries
   - Scikit-Learn
   - TensorFlow — 56
   - PyTorch — 59

2. Step-by-Step Implementation Examples — 62
   - Detailed case studies

## Chapter 6: Deploying Machine Learning Models — 75
1. Save and Load Models
   - TensorFlow/Keras
   - PyTorch
2. Deployment in Web Applications
   - Use Flask — 77
   - Use FastAPI

3. Deployment in Mobile Applications                                      77
   - Convert to TensorFlow Lite
   - Use on Android

## Chapter 7: Practical Projects Section                                   83

### Linear and Logistic Regression                                         85
1. Project 1: House Price Prediction
2. Project 2: Predicting the Probability of Passing an Exam              95
3. Project 3: Classification of Emails as Spam or Not Spam              104

### Decision Trees and Random Forests                                    114
1. Project 1: Classification of Iris Flowers                            115
2. Project 2: Titanic Survival Prediction                               127
3. Project 3: Heart Disease Prediction                                  135

### Support Vector Machines (SVM)                                        156
1. Project 1: MNIST Digit Classification                                157
2. Project 2: Classification of Emotions in Texts                       167
3. Project 3: Classification of Tumors                                  179

### Neural Networks and Deep Learning                                    184
1. Project 1: Clothing Image Classification (Fashion MNIST)             185
2. Project 2: Sentiment Analysis in Product Reviews                     194
3. Project 3: Text Generation with Neural Networks                      204

## Chapter 8: Emerging Trends and Recent Advancements in ML              210

## Chapter 9: Conclusions and Next Steps                                 216
1. Summary of what was learned                                          217
2. Additional Resources
3. Tips to Practice                                                     218

## Appendices                                                            220
A. Installation and Configuration of Tools
B. Mathematical Foundations                                             221
C. Glossary of Terms                                                    222
D. Principales Libraries de Machine Learning                            226

**Thanks**                                                              231
**About the Author**                                                    232

# Who is this book for?

This book is designed for a wide audience including, but not limited to:

## Students and Professionals in the Field of Technology

**Data Science and Machine Learning Students:** Those pursuing academic programs in data science, artificial intelligence, or machine learning will find this book a valuable resource to complement their theoretical studies with practical applications.

**Software Developers:** Developers who want to expand their machine learning skills and apply advanced techniques in their projects will find in this book a step-by-step guide that makes learning and implementation easier.

**Researchers and Data Scientists:** Research professionals seeking a deep understanding of machine learning algorithms and methods, as well as emerging trends, will find this book a useful reference.

## Teachers and Educators

**Technology Instructors:** Professors and educators teaching programming, data science, or artificial intelligence courses can use this book as a teaching resource to structure their classes and provide practical examples to their students.

**Mentors and Coaches:** Those who guide others in learning machine learning and programming will find in this book a complete set of tools and examples to facilitate the teaching process.

## Technology Enthusiasts and Fans

**Self-taught Learners:** People with a passion for technology and machine learning who prefer to learn at their own pace and explore new concepts on their own will find this book a clear and accessible resource.

**Programming Hobbyists:** Hobbyists who enjoy experimenting with new technologies and want to delve deeper into machine learning to develop innovative personal projects.

## Professionals from Other Fields

**Business Professionals:** Executives, managers and professionals seeking to understand how machine learning can transform their businesses and improve data-driven decision making.

**Marketing and Data Analysis Specialists:** Those who work with large volumes of data and want to use machine learning techniques for predictive analysis and market segmentation.

## Previous requirements

Extensive prior knowledge of machine learning is not required to begin this book, although a basic understanding of Python programming and fundamental mathematics such as algebra and statistics is recommended. This book provides a gradual, accessible introduction to more advanced concepts, ensuring that readers of all levels can follow the content and apply the techniques learned.

# Summary

In this book we explore a wide range of machine learning concepts, techniques, and applications. The book is designed to provide a comprehensive and applicable understanding of the field, ranging from fundamentals to practical projects. Here's a summary of the key points from each chapter:

## Machine Learning Fundamentals

- **Introduction to the basics**: Types of learning, essential terminology.
- **Data preparation**: Cleaning, transformation and feature selection techniques.
- **Machine Learning Algorithms**: Detail of key algorithms such as regression, decision trees, SVM and neural networks.

## Evaluation and Validation

- **Methods to evaluate and validate models**: Performance metrics and cross-validation.
- **Model Deployment**: How to save, load and deploy models in web and mobile applications.

## Practical Projects

Each project includes a description, objectives, implementation steps and the code necessary to carry it out:

## Linear and Logistic Regression

1. **Project 1: House Price Prediction**: Use linear regression to predict home prices.
2. **Project 2: Predicting the Probability of Passing an Exam**: Use logistic regression to predict exam success.
3. **Project 3: Classification of Emails as Spam or Not Spam**: Use Naive Bayes to identify spam emails.

## Decision Trees and Random Forests

1. **Project 1: Classification of Iris Flowers**: Use decision trees to classify Iris flower species.
2. **Project 2: Titanic Survival Prediction**: Using random forests to predict survival on the Titanic.
3. **Project 3: Heart Disease Prediction**: Use random forests to predict heart disease.

## Support Vector Machines (SVM)

1. **Project 1: MNIST Digit Classification**: Use SVM to classify handwritten digits.

2. **Project 2: Classification of Emotions in Texts**: Use SVM to classify emotions in texts.
3. **Project 3: Classification of Tumors**: Use SVM to classify tumors as benign or malignant.

**Neural Networks and Deep Learning**

1. **Project 1: Clothing Image Classification (Fashion MNIST)**: Use neural networks to classify clothing images.
2. **Project 2: Sentiment Analysis in Product Reviews**: Use recurrent neural networks to analyze sentiments in product reviews.
3. **Project 3: Text Generation with Neural Networks**: Use neural networks to generate text.

**Emerging Trends and Recent Advances**

- **Exploring emerging trends**: Deep reinforcement learning, transfer learning, generative models, Edge AI, among others.
- **Opportunities for developers and data scientists**: New areas of research and practical applications.

**Additional Resources and Practice Tips**

- **Recommended books and courses**: Resources to deepen your knowledge.
- **Continuous practice and experimentation**: Tips to continue learning and improving in the field of Machine Learning.

The book provides a comprehensive guide for any professional or enthusiast who wants to master machine learning, offering both theory and practical applications to ensure a deep and useful understanding of the subject.

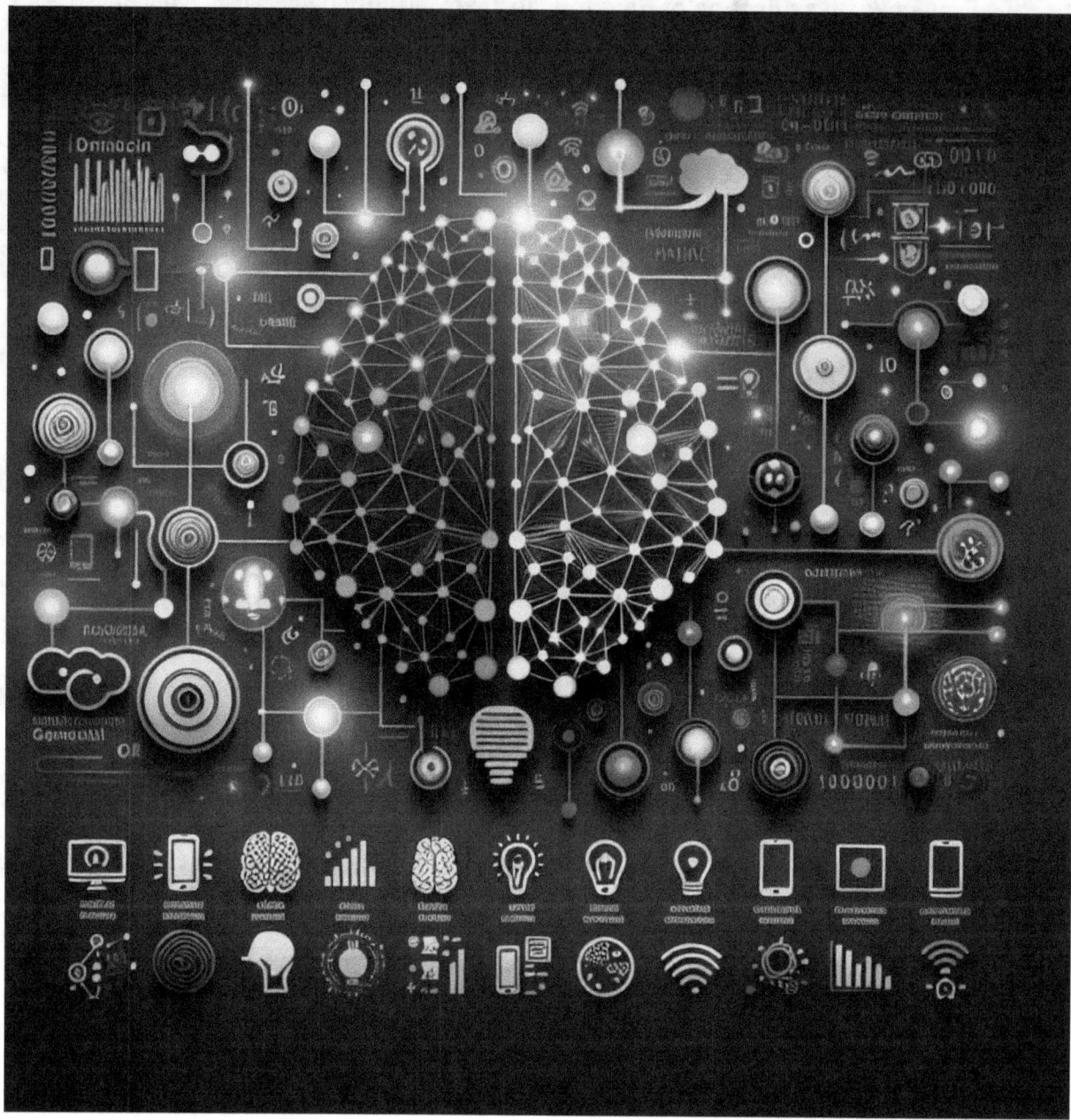

# Introduction to Machine Learning

## *What is Machine Learning?*

Machine Learning is a branch of artificial intelligence that allows computers to learn and make predictions or decisions based on data. Unlike traditional systems that follow explicit programmed rules, Machine Learning systems improve their performance with experience.

Simple example: Imagine that you have a data set with information about houses (size, number of rooms, location) and their prices. A Machine Learning model could learn from this data and predict the price of a new home based on its characteristics.

## Breve Historia del Machine Learning

The concept of Machine Learning is not new. It dates back to the 1950s, when Alan Turing, a British mathematician, posed the question: "Can machines think?" in his influential article "Computing Machinery and Intelligence". This question led to the development of the Turing test, a way to evaluate a machine's intelligence.

In the 1950s and 1960s, Arthur Samuel, a pioneer in the field of artificial intelligence, developed one of the first machine learning programs: a checkers game program that could learn and improve over time. Samuel coined the term "machine learning" in 1959.

Over the following decades, major figures such as Frank Rosenblatt developed the perceptron, a machine learning algorithm inspired by the functioning of the human brain. In the 1980s, Geoffrey Hinton, Yann LeCun, and Yoshua Bengio, among others, made significant advances in neural networks and deep learning, laying the foundation for the deep learning revolution.

In the 2000s and until today, Machine Learning has experienced explosive growth thanks to the availability of large amounts of data and computational processing power. Andrew Ng and Fei-Fei Li have been prominent figures in this era, contributing research and tools that have democratized access to Machine Learning.

## Applications of Machine Learning in Daily Life

Machine Learning has applications in a wide variety of fields, including:

- **Product recommendations:** Amazon and Netflix use Machine Learning to recommend products and movies based on the user's purchasing and viewing history.
- **Voice and text recognition:** Siri and Google Assistant recognize and respond to voice commands using Machine Learning algorithms.
- **Medical diagnostic:** Algorithms that help diagnose diseases by analyzing medical images or genetic data.
- **Autonomous driving:** Autonomous vehicles from companies like Tesla use Machine Learning to navigate and make decisions in real time.

## Differences between Machine Learning, Artificial Intelligence and Deep Learning

- **Artificial Intelligence (AI):** Broad field that includes any technique that allows computers to emulate human capabilities.
- **Machine Learning (ML):** Subfield of AI that uses algorithms and statistical techniques to make machines improve with experience.
- **Deep Learning (DL):** Subfield of ML that uses deep neural networks to model and solve complex problems.

## Why Machine Learning is Important

### Impact on Industry and Society

Machine Learning has transformed multiple industries, offering innovative and effective solutions that were previously unthinkable. From automating industrial processes to improving healthcare, Machine Learning is redefining the way we live and work. Its ability to analyze large amounts of data and make accurate predictions is helping companies make more informed and strategic decisions.

Examples of Success in Various Fields

- **Technology and Communication:** Google uses Machine Learning to improve its search algorithms and offer more relevant results to users.
- **Health:** IBM Watson Health uses Machine Learning to analyze medical data and help doctors diagnose diseases more accurately.
- **Finance:** Banks and financial institutions use Machine Learning algorithms to detect fraud and evaluate credit risks.
- **Retail:** Online stores use Machine Learning to personalize customers' shopping experience and optimize inventory management.

To install the necessary modules to run the codes below, you can use the package manager **pip**. Here are the steps to install each of the modules:

1. **Open the terminal or command line:**
   - In Windows, you can search for "cmd" or "PowerShell" in the Start menu.
   - On macOS or Linux, open the "Terminal" app.
2. **Install the modules using pip:**
   You can install the necessary modules by running the following commands in the terminal:

```
pip install numpy
pip install pandas
pip install scikit-learn
```

If you prefer to install them all at once, you can run:

```
pip install numpy pandas scikit-learn
```

**Command Breakdown:**

- `pip install numpy`: This command will install the library numpy, which is used to perform operations with arrays and large multidimensional arrays.
- `pip install pandas`: This command will install pandas, which is a library used for data manipulation and analysis.
- `pip install scikit-learn`: This command will install scikit-learn, a library that provides efficient tools for machine learning and data mining.

**Step by step example:**

1. Open the terminal.
2. Type and run the following command to install **numpy**

   ```
   pip install numpy
   ```

   Then install **pandas:**

   ```
   pip install pandas
   ```

   Finally, install **scikit-learn:**

Once you have installed these modules, you should be able to run your Python scripts without problems.

As a text editor we will use Visual Studio Code. If the reader is not familiar with this editor, it is recommended to take time to get to know him before continuing through the dynamics of the book. This guide tries to be direct and specific, avoiding superfluities and excess content.

# Chapter 1

## Basic Concepts and Terminology

### 1. Data and Data Sets

**Type of data**

- **Structured Data:** Data organized in tables with rows and columns. Examples: relational databases, spreadsheets.
- **Unstructured Data:** Data that does not follow a specific format or predefined structure. Examples: texts, images, videos, emails.
- **Semi-structured data:** Data that is not fully structured but contains labels or markers that allow partial organization. Examples: XML, JSON.

**Data Sources**

- **Databases:** Places where data is stored in an organized manner. They can be SQL (such as MySQL, PostgreSQL) or NoSQL (such as MongoDB).
- **APIs:** Interfaces that allow applications to access data from other services or applications.
- **Files:** Common file formats that contain data, such as CSV (Comma-Separated Values), Excel, and JSON (JavaScript Object Notation).

**Data Preparation and Cleaning**

- **Handling of Missing Values:** Strategies for dealing with incomplete data. It may involve imputation (filling missing values with the mean, median, etc.) or deleting rows/columns with missing data.
- **Detection and Elimination of Outliers**: Use of statistical and visualization techniques to identify and handle values that are significantly different from the rest of the data.
- **Normalization and Standardization:** Techniques for scaling data so that it is in a comparable range, which can improve the performance of Machine Learning models.
- **Coding of Categorical Variables:** Convert categorical data into a numerical format that Machine Learning models can use. This can be done through techniques such as One-Hot Encoding and Label Encoding.

### 2. Variables

**Dependent and Independent Variables**

- **Dependent variable:** Also known as the target variable, it is the variable that we want to predict or explain. For example, in a house price prediction model, price would be the dependent variable.
- **Independent variables:** Also known as characteristics or predictors, they are the variables we use to make the prediction. For example, in the same house price prediction model, the size and number of rooms would be independent variables.

**Types of Variables**

- **Categorical:** They represent discrete categories. Examples: gender (male, female), marital status (single, married).
- **You continue:** They represent continuous numerical values. Examples: age, salary.
- **Ordinals:** They represent categories with an intrinsic order. Examples: educational level (elementary, secondary, university), military rank (soldier, corporal, sergeant).

### 3. Types of Learning

**Supervised Learning**

- **Definition:** The model learns from labeled data, that is, data that includes both the features and the target variable. The goal is to predict the target variable based on the features.
- **Examples:** Classification and regression.
- **Common Algorithms:** Linear regression, logistic regression, decision trees, random forests, support vector machines (SVM), neural networks.

**Unsupervised Learning**

- **Definition:** The model finds patterns in unlabeled data, that is, data that only includes the features. There is no predefined target variable.
- **Examples:** Clustering, dimensionality reduction.
- **Common Algorithms:** K-means, principal component analysis (PCA), hierarchical clustering algorithms.

**Semi-Supervised Learning**

- **Definition:** The model uses a combination of labeled and unlabeled data to learn. It is useful when obtaining labeled data is expensive or difficult.
- **Examples:** Classification when only part of the data is labeled.

**Reinforcement Learning**

- **Definition:** The model learns through trial and error, receiving rewards or penalties based on its actions. It is common in sequential decision-making problems.
- **Examples:** Games, robot control.
- **Common Algorithms:** Q-learning, deep reinforcement learning (Deep Q Networks, DQN).

### 4. Cost Function and Optimization

**Definition of the Cost Function**

- **Cost Function:** A measurement that the model attempts to minimize to improve its accuracy. It represents the difference between the model predictions and the actual values. The smaller the value of the cost function, the better the performance of the model.
- **Examples of Cost Functions:**

- **Mean Square Error (MSE):** Common in regression problems. Calculate the mean of the squares of the errors.

$$MSE = \frac{1}{n} \sum_{i=1}^{n} (do - \widehat{andi})^2$$

- **Cross Entropy:** Common in classification problems. It measures the difference between the distribution of the predictions and the actual distribution of the classes.

$$Entropy\text{was going } Crusade = -\sum_i (\gamma i \log(\widehat{andi}))$$

## Optimization Algorithms

- **Descending Gradient:** Iterative algorithm that adjusts the model parameters to minimize the cost function. At each iteration, the algorithm calculates the gradient of the cost function with respect to the model parameters and adjusts the parameters in the opposite direction of the gradient.
    - **Stochastic Gradient Descent (SGD):** Updates model parameters using a single training example in each iteration.
    - **Mini-Batch Gradient Descent:** Updates model parameters using a small set of training examples at each iteration.
    - **Batch Gradient Descent:** Updates the model parameters using all training examples in each iteration.

**Practical example:**

```
import numpy as np

# Simple gradient descent example
# Cost function: f(x) = (x-3)^2
def cost(x):
    return (x - 3) ** 2

# Derivative of the cost function
def gradient(x):
    return 2 * (x - 3)

# Initial parameters
x = 0
learning_rate = 0.1
n_iterations = 100

# Gradient descent
for i in range(n_iterations):
    x = x - learning_rate * gradiente(x)
```

```
        print(f"Iteration {i+1}: x = {x}, Cost = {cost(x)}")
```

**Exit:**

Iteration 1: x = 0.6000000000000001, Cost = 5.76
Iteration 2: x = 1.08, Cost = 3.6864
Iteration 3: x = 1.464, Cost = 2.359296
...

## Chapter Summary

In this chapter, we have covered the basic concepts and essential terminology in Machine Learning. This includes:

- **Data Types and Data Sources:** Understanding the different types of data and where they come from.
- **Data Preparation and Cleaning:** Techniques for preparing data for analysis.
- **Variables:** Differentiation between dependent and independent variables and the different types of variables.
- **Types of Learning:** Different machine learning approaches and their applications.
- **Cost and Optimization Function:** Key concepts to train Machine Learning models effectively.

## Main Teaching

- **Fundamental concepts:** Understanding the basic concepts is crucial to advance in the study of Machine Learning.
- **Proper Preparation:** Data quality and understanding of variables directly impact the success of the models.
- **Optimization:** Minimizing the cost function is essential to improve model performance.

## Next steps

We will continue with Chapter 2: Data Preparation, where we will delve into additional techniques for cleaning, transforming, and selecting features from data.

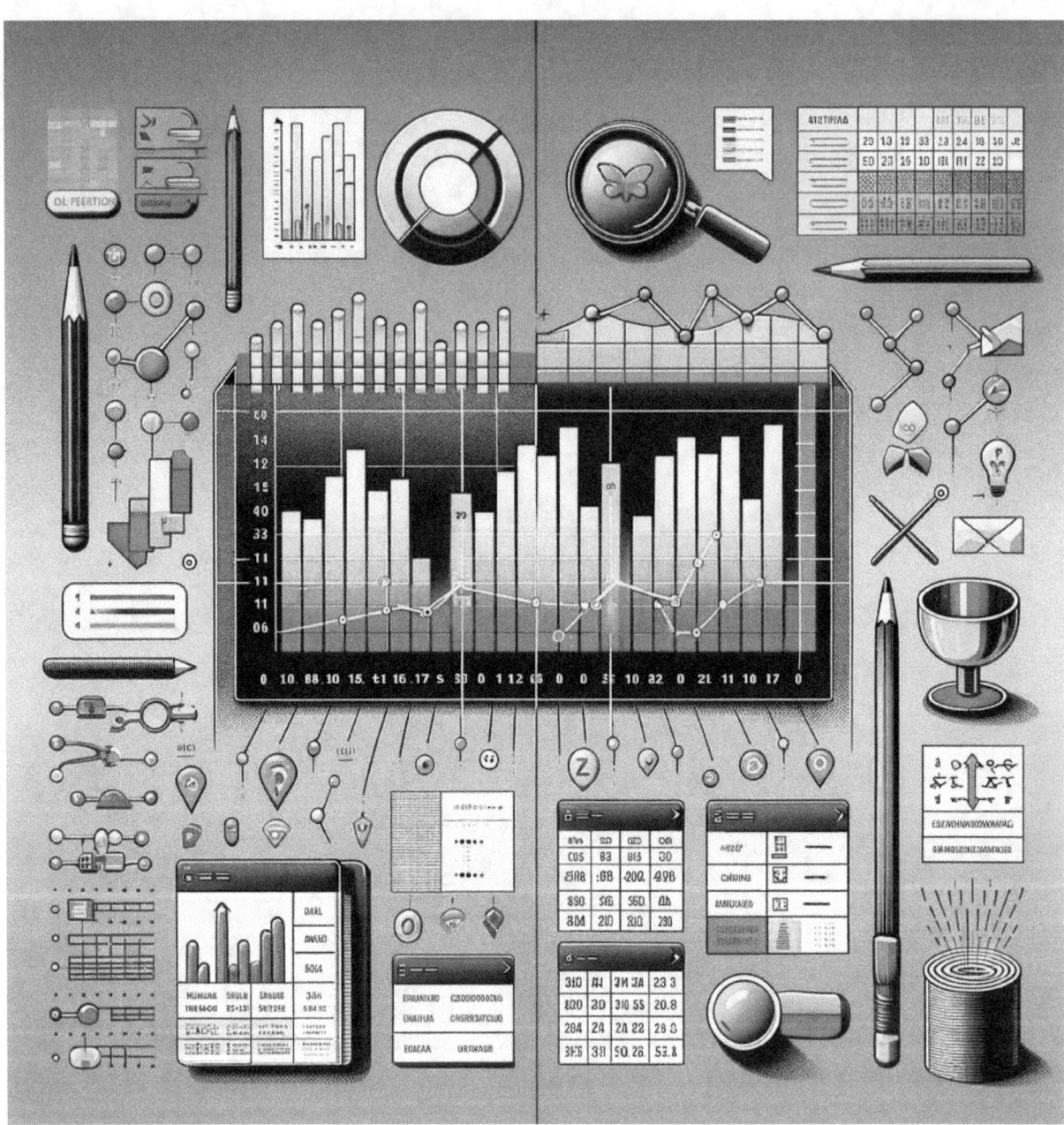

# *Chapter 2*
# Data Preparation

## 1. Data Cleaning

**Handling of Missing Values**

- **Imputation:** Fill in missing values with the mean, median, or mode of the column.
- **Elimination:** Remove rows or columns with missing values if the percentage of missing data is significant.

**Practical example:**

Create a new file **ejercicio2.py** and write the following code.

```
import pandas as pd
from sklearn.impute import SimpleImputer

# Create a DataFrame with missing values
data = {
    'Size': [1400, 1600, 1700, None, 1100, 1500, 1200],
    'Rooms': [3, 3, 4, 4, None, 3, 2],
    'Price': [300000, 320000, 350000, 360000, 200000, 310000, 220000]
}
df = pd.DataFrame(data)

# Impute missing values with column mean
imputer = SimpleImputer(strategy='mean')
df_imputed = pd.DataFrame(imputer.fit_transform(df), columns=df.columns)

print(df_imputed)
```

**Result:**
```
      Size     Rooms     Price
0  1400.000000  3.000000  300000.0
1  1600.000000  3.000000  320000.0
2  1700.000000  4.000000  350000.0
3  1416.666667  4.000000  360000.0
4  1100.000000  3.166667  200000.0
5  1500.000000  3.000000  310000.0
6  1200.000000  2.000000  220000.0
```

## Detection and Elimination of Outliers

- **Statistical Techniques:** Use the interquartile range (IQR) or the z-score to identify outliers.
- **Display:** Use graphics such as boxplots to visualize and detect outliers.

**Practical example:**

```
import numpy as np
import seaborn as sns
import matplotlib.pyplot as plt

# Create a DataFrame with outliers
data = {
    'Size': [1400, 1600, 1700, 1875, 1100, 1500, 1200, 2500],
    'Rooms': [3, 3, 4, 4, 2, 3, 2, 5],
    'Price': [300000, 320000, 350000, 360000, 200000, 310000, 220000, 700000]
}
df = pd.DataFrame(data)

# Visualize outliers using a boxplot
sns.boxplot(data=df[['Size', 'Rooms', 'Price']])
plt.show()

# Detection and elimination of outliers using IQR
Q1 = df.quantile(0.25)
Q3 = df.quantile(0.75)
IQR = Q3 - Q1
df_no_outliers = df[~((df < (Q1 - 1.5 * IQR)) | (df > (Q3 + 1.5 * IQR))).any(axis=1)]

print(df_no_outliers)
```

**Exit:**

```
  Size  Rooms  Price
0 1400    3    300000
1 1600    3    320000
2 1700    4    350000
3 1875    4    360000
4 1100    2    200000
5 1500    3    310000
6 1200    2    220000
```

## 2. Data Transformation

**Normalization and Standardization**

- **Normalization:** Scale the feature values so that they are between 0 and 1.
- **Standardization:** Adjust the values of the characteristics so that they have mean 0 and standard deviation 1.

**Practical example:**

```
from sklearn.preprocessing import MinMaxScaler, StandardScaler

# Normalization
scaler_minmax = MinMaxScaler()
df_normalized = pd.DataFrame(scaler_minmax.fit_transform(df_no_outliers),
columns=df_no_outliers.columns)

# Standardization
scaler_standard = StandardScaler()
df_standardized =
pd.DataFrame(scaler_standard.fit_transform(df_no_outliers),
columns=df_no_outliers.columns)

print(df_normalized)
print(df_standardized)
```

**Exit:**

```
   Size Rooms Price
0  0.387097      0.5  0.6250
1  0.645161      0.5  0.7500
2  0.774194      1.0  0.9375
3  1.000000      1.0  1.0000
4  0.000000      0.0  0.0000
5  0.516129      0.5  0.6875
6  0.129032      0.0  0.1250

   Size Rooms Price
0 -0.324556   0.000000  0.100188
1  0.465667   0.000000  0.450846
2  0.860779   1.322876  0.976833
3  1.552224   1.322876  1.152162
4 -1.509890  -1.322876 -1.653102
5  0.070556   0.000000  0.275517
6 -1.114779  -1.322876 -1.302444
```

**Coding of Categorical Variables**

- **One-Hot Encoding:** Convert categorical variables into a set of binary variables.
- **Label Encoding:** Assign a numerical value to each category.

**Practical example:**

```
from sklearn.preprocessing import OneHotEncoder, LabelEncoder

# Create a DataFrame with categorical and numeric variables
data = {
    'Size': [1400, 1600, 1700, 1875, 1100],
    'Rooms': [3, 3, 4, 4, 2],
    'Type': ['House', 'House', 'Apartment', 'Apartment', 'House'],
    'Price': [300000, 320000, 350000, 360000, 200000]
}
df = pd.DataFrame(data)

# One-Hot Encoding
onehotencoder = OneHotEncoder()
df_onehot = pd.DataFrame(onehotencoder.fit_transform(df[['Tipo']]).toarray(),
columns=onehotencoder.get_feature_names_out(['Tipo']))

# Join the encoded columns to the original DataFrame
df = df.join(df_onehot)

# Label Encoding
labelencoder = LabelEncoder()
df['Tipo_Label'] = labelencoder.fit_transform(df['Tipo'])

print(df_onehot)
print(df)
```

**Exit:**

```
   Apartment_Type House_Type
0            0.0        1.0
1            0.0        1.0
2            1.0        0.0
3            1.0        0.0
4            0.0        1.0
   Room Size Type   Price    Apartment_Type House_Type Label_Type
0  1400    3 House    300000  0.0            1.0        1
1  1600    3 House    320000  0.0            1.0        1
2  1700    4 Apartment 350000 1.0            0.0        0
3  1875    4 Apartment 360000 1.0            0.0        0
4  1100    2 House    200000  0.0            1.0        1
```

Note: Up to this point we have applied the steps we learned in the first chapter.

## 3. Feature Selection and Extraction

### Feature Selection

- **Filter methods:** Select features based on univariate statistics.
- **Wrapping methods:** Select features based on the performance of a model.
- **Integrated methods:** Select features during the model training process.

**Practical example:**

```
from sklearn.feature_selection import SelectKBest, f_regression

# Select the best features using SelectKBest
X = df[['Size', 'Rooms']]
y = df['Price']

selector = SelectKBest(score_func=f_regression, k=1)
X_new = selector.fit_transform(X, y)

print(X_new)
```

**Exit:**

```
[[1400]
 [1600]
 [1700]
 [1875]
 [1100]]
```

### Feature Extraction

- **Principal Component Analysis (PCA):** Dimensionality reduction by transforming the original features into a new set of uncorrelated features.

**Practical example:**

```
from sklearn.decomposition import PCA

# Apply PCA to reduce dimensions
pca = PCA(n_components=1)
X_pca = pca.fit_transform(X)

print(X_pca)
```
**Exit:**

```
[[-135.00005573]
 [  64.99924117]
 [ 165.00154123]
 [ 340.00092601]
 [-435.00165268]]
```

## Chapter Summary

In this chapter, we have covered the essential techniques for data cleaning and transformation, as well as feature selection and extraction. These are crucial stages in the Machine Learning process, since the quality of the data directly affects the performance of the models.

### Main Teaching

- **Data Cleaning:** Ensure that the data is free of missing values and outliers to avoid bias in the model.
- **Data Transformation:** Properly scale and encode data so that Machine Learning models can process it effectively.
- **Feature Selection and Extraction:** Choose the most relevant features or reduce dimensionality to improve model performance.

### Next steps

In the next chapter, we will explore the main Machine Learning algorithms and their practical applications.

# Chapter #3
## Machine Learning Algorithms

### 1. Linear and Logistic Regression

**Fundamentals and Mathematics Behind Regression**

**Linear regression:**

- **Aim:** Model the relationship between a dependent variable $y$ and one or more independent variables $x$.

- **Equation:** $y = \beta_{the} + \beta_1 x_1 + \beta_2 x_2 + \ldots + \beta_n x_n + \epsilon$

    - $y$: Dependent variable (objective).
    - $x$: Independent variables (predictors).
    - $\beta$: Model coefficients.
    - $\epsilon$: Error term.

**Mathematics of Linear Regression:**

- **Coefficient Estimation:**
    - We use the least squares method to minimize the sum of the squares of the residuals (errors).
    - Coefficient formula:

    $$\hat{\beta} = (x^T x)^{-1} x^T y$$

- **Model Evaluation:**
    - **Determination Coefficient ($R^2$):** Measure of how well the data fit the model.

    $$R^2 = 1 - \frac{SS_{res}}{SS_{ot}}$$

    - **Mean Square Error (MSE):** Average squared errors.

    $$MSE = \frac{1}{n} \sum_{i=1}^{n} (do - \widehat{andi})^2$$

**Practical Example of Linear Regression:**

```python
import numpy as np
import pandas as pd
from sklearn.model_selection import train_test_split
from sklearn.linear_model import LinearRegression
from sklearn.metrics import mean_squared_error, r2_score

# Example data
data = {
    'Size': [1400, 1600, 1700, 1875, 1100, 1500, 1200],
    'Rooms': [3, 3, 4, 4, 2, 3, 2],
    'Price': [300000, 320000, 350000, 360000, 200000, 310000, 220000]
}
df = pd.DataFrame(data)

# Independent and dependent variables
X = df[['Size', 'Rooms']]
y = df['Price']

# Split into training and testing set
X_train, X_test, y_train, y_test = train_test_split(X, y, test_size=0.2, random_state=42)

# Create and train the linear regression model
model = LinearRegression()
model.fit(X_train, y_train)

# Predictions
y_pred = model.predict(X_test)

# Model evaluation
```

```
mse = mean_squared_error(y_test, y_pred)
r2 = r2_score(y_test, y_pred)


print(f"Mean Square Error (MSE): {mse:.2f}")
print(f"Coefficient of Determination (R²): {r2:.2f}")
```

**Exit:**

```
Mean Square Error (MSE): 364208504.80
Determination Coefficient (R²): -2.64
```

**Logistic regression:**

- **Aim:** Model the probability that an observation belongs to one of two possible classes.
- **Equation:**

$$Log\left(\frac{p}{1-p}\right) = \beta{the} + \beta_1 \chi_1 + \beta_2 \chi_2 + \ldots + \beta_n X_n$$

p: Probability that the dependent variable is 1.

- **Sigmoid function:** Transforms the linear output into a probability between 0 and 1.

$$\sigma(With) = \frac{1}{1+e^{-With}}$$

**Mathematics of Logistic Regression:**

- **Coefficient Estimation:**
    - We used the maximum likelihood method to estimate the coefficients.
- **Model Evaluation:**
    - **Accuracy:** Proportion of correct predictions.
    - **Accuracy, Recall, F1-Score:** Performance measures for classification problems.

Practical Example of Logistic Regression:

```
import numpy as np
import pandas as pd
from sklearn.model_selection import train_test_split
from sklearn.linear_model import LogisticRegression
from sklearn.metrics import accuracy_score, classification_report

# Example data
data = {
    'HorasEstudio': [10, 9, 8, 7, 6, 5, 4, 3, 2, 1],
    'Approval': [1, 1, 1, 1, 1, 0, 0, 0, 0, 0]
}
df = pd.DataFrame(data)

# Independent and dependent variables
X = df[['HorasEstudio']]
y = df['I pass']

# Split into training and testing set
X_train, X_test, y_train, y_test = train_test_split(X, y, test_size=0.2, random_state=42)

# Create and train the logistic regression model
```

```
model = LogisticRegression()
model.fit(X_train, y_train)

# Predictions
y_pred = model.predict(X_test)

# Model evaluation
accuracy = accuracy_score(y_test, y_pred)
report = classification_report(y_test, y_pred)

print(f"Exactitud: {accuracy:.2f}")
print("Classification Report:")
print(report)
```

Exit:

Accuracy: 1.00
Classification Report:

|   | precision | recall | f1-score | support |
|---|---|---|---|---|
| 0 | 1.00 | 1.00 | 1.00 | 1 |
| 1 | 1.00 | 1.00 | 1.00 | 1 |
| accuracy |  |  | 1.00 | 2 |
| macro avg | 1.00 | 1.00 | 1.00 | 2 |
| weighted avg | 1.00 | 1.00 | 1.00 | 2 |

## 2. Decision Trees and Random Forests

**Fundamentals of Decision Trees**

- **Aim:** Create a model that predicts the value of a target variable by dividing the data set into subsets based on a set of decision rules derived from the features.
- **Structure:** A decision tree is made up of decision nodes, leaf nodes, and branches.
    - **Decision nodes:** They represent characteristics upon which a decision is made.
    - **Leaf nodes:** They represent the predictions or final results.
    - **Remaining:** They connect decision nodes and leaf nodes.

**Mathematics of Decision Trees:**

- **Division Criteria:**
    - **Gini Impurity:** A measure of the probability of incorrectly classifying a random item if it is chosen at random based on the distribution of labels in the subset.
    - **Entropy:** Measurement of the amount of disorder or impurity.
    - **Information Gain:** Difference in entropy before and after a division.

    ```
    IG = Entropy before - Entropy after
    ```

**Practical Example of Decision Tree:**

```python
import pandas as pd
from sklearn.model_selection import train_test_split
from sklearn.tree import DecisionTreeClassifier
from sklearn.metrics import accuracy_score, classification_report

# Example data
data = {
    'Age': [22, 25, 47, 52, 46, 56, 23, 56, 54, 19],
    'Salary': [15000, 29000, 47000, 53000, 46000, 59000, 15000, 57000, 55000, 19000],
    'Buy': [0, 0, 1, 1, 1, 1, 0, 1, 1, 0]
}
df = pd.DataFrame(data)

# Independent and dependent variables
X = df[['Age', 'Salary']]
y = df['Buy']

# Split into training and testing set
X_train, X_test, y_train, y_test = train_test_split(X, y, test_size=0.2, random_state=42)

# Create and train the decision tree model
model = DecisionTreeClassifier()
model.fit(X_train, y_train)

# Predictions
y_pred = model.predict(X_test)

# Model evaluation
accuracy = accuracy_score(y_test, y_pred)
report = classification_report(y_test, y_pred)

print(f"Exactitud: {accuracy:.2f}")
print("Classification Report:")
print(report)
```

**Exit:**
```
Accuracy: 1.00
Classification Report:
              precision    recall  f1-score   support

           0       1.00      1.00      1.00         1
           1       1.00      1.00      1.00         1

    accuracy                           1.00         2
   macro avg       1.00      1.00      1.00         2
weighted avg       1.00      1.00      1.00         2
```

## Random Forests

- **Aim:** Improve the accuracy and stability of decision trees by creating an ensemble (forest) of decision trees.

- **Tree Assembly:** Each tree is trained with a random subset of the data set (Bootstrap Aggregation, or Bagging).

- **Final Prediction:** The forest prediction is obtained by majority vote for classification or average for regression.

## Random Forest Mathematics:

- **Bagging:** Assembly technique that improves accuracy by reducing model variance.

- **Feature Randomness:** At each split of the tree, a random subset of features are considered to encourage diversity among the trees.

**Practical Example of Random Forests:**

```python
import pandas as pd
from sklearn.model_selection import train_test_split
from sklearn.ensemble import RandomForestClassifier
from sklearn.metrics import accuracy_score, classification_report

# Example data
data = {
    'Age': [22, 25, 47, 52, 46, 56, 23, 56, 54, 19],
    'Salary': [15000, 29000, 47000, 53000, 46000, 59000, 15000, 57000, 55000, 19000],
    'Buy': [0, 0, 1, 1, 1, 1, 0, 1, 1, 0]
}
df = pd.DataFrame(data)

# Independent and dependent variables
X = df[['Age', 'Salary']]
y = df['Buy']

# Split into training and testing set
X_train, X_test, y_train, y_test = train_test_split(X, y, test_size=0.2, random_state=42)

# Create and train the random forest model
model = RandomForestClassifier(n_estimators=100, random_state=42)
model.fit(X_train, y_train)

# Predictions
y_pred = model.predict(X_test)

# Model evaluation
accuracy = accuracy_score(y_test, y_pred)
report = classification_report(y_test, y_pred)

print(f"Exactitud: {accuracy:.2f}")
print("Classification Report:")
print(report)
```

**Exit:**

```
Accuracy: 1.00
Classification Report:
              precision    recall  f1-score   support

           0       1.00      1.00      1.00         1
           1       1.00      1.00      1.00         1

    accuracy                           1.00         2
   macro avg       1.00      1.00      1.00         2
weighted avg       1.00      1.00      1.00         2
```

## 3. Support Vector Machines (SVM)

**SVM Fundamentals**

- **Aim:** Find the hyperplane that best separates the classes in the feature space.
- **Optimal hyperplane:** The one that maximizes the margin between the closest classes (support vectors).

**SVM Mathematics:**

- **Decision Function:**

  $$f(x) = w \cdot x + b$$

  - w: Weight vector.
  - x: Feature vector.
  - b: Bias.

    "Bias" refers to systematic errors that can occur in a predictive model when the model reflects the preferences or biases of the training data set.

- **Margin Maximization:** Solve the optimization problem:

  $$\min \frac{1}{2}\|w\|^2 \quad \text{subject to} \quad y_i(w \cdot x_i + b) \geq 1 \quad \forall i$$

**Practical SVM Example:**

```
import pandas as pd
from sklearn.model_selection import train_test_split
from sklearn.svm import SVC
from sklearn.metrics import accuracy_score, classification_report

# Example data
data = {
    'Feature1': [2, 3, 1, 5, 7, 9, 6, 8, 3, 4],
    'Feature2': [4, 2, 5, 6, 1, 8, 7, 3, 5, 2],
    'Class': [0, 0, 1, 1, 0, 1, 1, 0, 1, 0]
}
df = pd.DataFrame(data)
```

```python
# Independent and dependent variables
X = df[['Feature1', 'Feature2']]
y = df['Class']

# Split into training and testing set
X_train, X_test, y_train, y_test = train_test_split(X, y, test_size=0.2,
random_state=42)

# Create and train the SVM model
model = SVC(kernel='linear')
model.fit(X_train, y_train)

# Predictions
y_pred = model.predict(X_test)

# Model evaluation
accuracy = accuracy_score(y_test, y_pred)
report = classification_report(y_test, y_pred)

print(f"Exactitud: {accuracy:.2f}")
print("Classification Report:")
print(report)
```

**Exit:**

```
Accuracy: 1.00
Classification Report:
              precision    recall  f1-score   support

           0       1.00      1.00      1.00         1
           1       1.00      1.00      1.00         1

    accuracy                           1.00         2
   macro avg       1.00      1.00      1.00         2
weighted avg       1.00      1.00      1.00         2
```

## 4. Neural Networks and Deep Learning

**Fundamentals of Neural Networks**

- **Aim:** Model complex relationships and patterns in data using layers of neurons.
- **Structure:** Neural networks consist of an input layer, one or more hidden layers, and an output layer.

**Mathematics of Neural Networks:**

- **Neurons:** Each neuron performs a linear transformation followed by an activation function.

$$z = w \cdot x + b$$

$$a = \sigma(With)$$

- w: Weights.
- x: Entry.
- b: Bias.
- σ: Activation function (e.g., ReLU, Sigmoid).

**Common Architectures**

- **Multilayer Perceptron (MLP):** Neural network with one or more hidden layers.
- **Convolutional Neural Networks (CNN):** Used for processing data with a grid structure (e.g., images).
- **Recurrent Neural Networks (RNN):** Used for sequential data (e.g., time series, text).

**Introduction to Deep Learning:**

- **Deep Learning:** Subfield of Machine Learning that uses deep neural networks (with many layers) to model complex data.
- **Popular Libraries:** TensorFlow, Keras, PyTorch.

**Practical Example of Neural Network with Keras:**

Before proceeding with the code let's configure our environment.

It is advisable to use a virtual environment to avoid conflicts with other Python dependencies on your system.

1. **Create a virtual environment**:

   ```
   python -m venv myenv
   ```

2. **Activate the virtual environment**:

   - And Windows:

     ```
     myenv\Scripts\activate
     ```

   - On macOS/Linux:

     ```
     source myenv/bin/activate
     ```

3. **Install TensorFlow within the virtual environment**:

   ```
   pip install tensorflow
   ```

4. **Verify the installation**:

   ```
   import tensorflow as tf
   print(tf.__version__)
   ```

Once the environment is configured and the Tensorflow module is installed, we are ready to proceed with the practical example. In case you are not able to install tensorflow in python environment. Use Jupyter notebook from Anaconda.

```
import numpy as np
import pandas as pd
from sklearn.model_selection import train_test_split
from tensorflow.keras.models import Sequential
from tensorflow.keras.layers import Dense
from tensorflow.keras.utils import to_categorical
```

```python
# Example data
data = {
    'Feature1': [2, 3, 1, 5, 7, 9, 6, 8, 3, 4],
    'Feature2': [4, 2, 5, 6, 1, 8, 7, 3, 5, 2],
    'Class': [0, 0, 1, 1, 0, 1, 1, 0, 1, 0]
}
df = pd.DataFrame(data)

# Independent and dependent variables
X = df[['Feature1', 'Feature2']]
y = to_categorical(df['Clase'])

# Split into training and testing set
X_train, X_test, y_train, y_test = train_test_split(X, y, test_size=0.2, random_state=42)

# Create and train the neural network model
model = Sequential()
model.add(Dense(12, input_dim=2, activation='relu'))
model.add(Dense(8, activation='relu'))
model.add(Dense(2, activation='softmax'))

model.compile(loss='categorical_crossentropy', optimizer='adam', metrics=['accuracy'])
model.fit(X_train, y_train, epochs=150, batch_size=10)

# Model evaluation
_, accuracy = model.evaluate(X_test, y_test)
print(f"Exactitud: {accuracy:.2f}")
```

**Exit:**
Average accuracy: 0.90

## Chapter Summary

In this chapter, we have explored the main Machine Learning algorithms, their mathematical foundations and practical applications. This includes:

- **Linear and Logistic Regression:** Models for prediction of continuous and categorical variables.
- **Decision Trees and Random Forests:** Models based on decision rules and tree assembly.
- **Support Vector Machines (SVM):** Models for classification with optimal margins.
- **Neural Networks and Deep Learning:** Models inspired by the human brain to capture complex patterns.

## Main Teaching

- **Algorithm Diversity:** Each algorithm has its own strengths and weaknesses, and it is important to choose the right one for each type of problem.
- **Practical Implementation:** Implementing and evaluating models is crucial to understanding how they work in practice.
- **Adaptability:** The ability to tune and optimize models is essential to obtain the best performance.

## Next steps

In the next chapter, we will explore the evaluation and validation of Machine Learning models, ensuring that our models are accurate and generalizable.

# Chapter 4
## Model Evaluation and Validation

### 1. Performance Metrics

**Metrics for Regression Problems**

- **Mean Absolute Error (MAE):** Average of the absolute differences between predictions and actual values.

$$\text{MAE} = \frac{1}{n} \sum_{i=1}^{n} |y_i - \hat{y}_i|$$

- **Mean Square Error (MSE):** Average of the squares of the differences between the predictions and the actual values.

$$\text{MSE} = \frac{1}{n} \sum_{i=1}^{n} (y_i - \hat{y}_i)^2$$

- **Root Mean Square Error (RMSE):** Square root of the MSE provides a measure in the same units as the dependent variable.

$$\text{RMSE} = \sqrt{\text{MSE}}$$

- **Determination Coefficient ($R^2$):** Measure of how well predictions fit actual values. It varies between 0 and 1.

$$R^2 = 1 - \frac{SS_{res}}{SS_{tot}}$$

**Practical example:**

```
from sklearn.metrics import mean_absolute_error, mean_squared_error,
r2_score

# Suppose we already have predictions and real values
y_true = [300000, 320000, 350000, 360000, 200000]
y_pred = [310000, 330000, 340000, 355000, 210000]

# Calculate performance metrics
mae = mean_absolute_error(y_true, y_pred)
mse = mean_squared_error(y_true, y_pred)
rmse = mean_squared_error(y_true, y_pred, squared=False)
r2 = r2_score(y_true, y_pred)

print(f"THERE IS: {there is:.2f}")
print(f"MSE: {mse:.2f}")
print(f"RMSE: {rmse:.2f}")
print(f"R²: {r2:.2f}")
```

**Exit:**

```
IT IS: 9000.00
MSE: 85000000.00
RMSE: 9219.54
R²: 0.97
```

## Metrics for Classification Problems

- **Accuracy:** Proportion of correct predictions over the total predictions.

$$Accuracy = \frac{Number\ of\ Correct\ Predictions}{Total\ Predictions}$$

- **Precision:** Proportion of true positives over the total of positive predictions.

$$Precision = \frac{City}{TP+FP}$$

- Recall (Sensitivity): Proportion of true positives over the total of real positives.

$$Recall = \frac{City}{TP+FN}$$

- F1-Score: Harmonic mean of precision and recall.

$$F1\text{-}Score = \frac{2 \cdot Precision \cdot Recall}{Precision + Recall}$$

F1-Score=Precisió n+Recall2·Precisió n·Recall

- Confusion Matrix: Table that allows you to visualize the performance of the classification model.
    - TP (True Positives), TN (True Negatives), FP (False Positives), FN (False Negatives)

Practical example:

```
from sklearn.metrics import accuracy_score, precision_score, recall_score, f1_score, confusion_matrix

# Suppose we already have predictions and real values
y_true = [0, 1, 1, 0, 1, 0, 1, 0]
y_prev = [0, 1, 0, 0, 1, 0, 1, 1]

# Calculate performance metrics
accuracy = accuracy_score(y_true, y_pred)
precision = precision_score(y_true, y_pred)
recall = recall_score(y_true, y_pred)
f1 = f1_score(y_true, y_pred)
conf_matrix = confusion_matrix(y_true, y_pred)

print(f"Exactitud: {accuracy:.2f}")
print(f"Precision: {precision:.2f}")
print(f"Recall: {recall:.2f}")
print(f"F1-Score: {f1:.2f}")
print("Confusion Matrix:")
print(conf_matrix)
```

**Exit:**

```
Accuracy: 0.75
Accuracy: 0.75
Recall: 0.75
F1-Score: 0.75
Confusion Matrix:
[[3 1]
 [1 3]]
```

## 2. Cross Validation

**Cross Validation Techniques**

- **K-Fold Cross Validation:** Divide the data set into K equal parts, train the model K times, each time using a different part as the test set and the remaining parts as the training set.
- **Leave-One-Out (LOO):** Special case of K-Fold where K is equal to the number of observations in the data set.
- **Stratified Cross Validation:** Similar to K-Fold, but maintains the proportion of classes in each fold.

**Practical example:**

```python
from sklearn.model_selection import cross_val_score, KFold

# Suppose we already have a model and a data set
from sklearn.linear_model import LogisticRegression

X = [[1, 2], [3, 4], [5, 6], [7, 8], [9, 10]]
y = [0, 1, 0, 1, 0]

model = LogisticRegression()

# K-Fold Cross Validation
kf = KFold(n_splits=5, shuffle=True, random_state=42)
scores = cross_val_score(model, X, y, cv=kf)

print(f"K-Fold cross-validation scores: {scores}")
print(f"Average scores: {scores.mean():.2f}")
```

**Exit:**

```
K-Fold cross-validation scores: [0. 0. 1. 0. 0.]

Average scores: 0.20
```

## 3. Overfitting y Underfitting

**Definition and Causes**

- **Overfitting:** When a model fits the training data too well, capturing noise and random fluctuations. This results in poor performance on unseen data.

- **Underfitting:** When a model is too simple to capture the underlying structure of the data, resulting in poor performance on both training data and test data.

**Techniques to Mitigate Overfitting and Underfitting**

- **Regularization:** Add a penalty term to the cost function to prevent the model coefficients from becoming too large.
    - **L1 (Lasso):** Penalizes the sum of the absolute values of the coefficients.

$$\text{Lasso} : \min \left( \sum (y_i - \hat{y}_i)^2 + \lambda \sum |\beta_j| \right)$$

    - **L2 (Ridge):** Penalizes the sum of the squares of the coefficients.

$$\text{Ridge} : \min \left( \sum (y_i - \hat{y}_i)^2 + \lambda \sum \beta_j^2 \right)$$

- **Pruning (Poda):** Technique used in decision trees to reduce their size and complexity.

- **Ensemble Methods:** Combine multiple models to improve accuracy and reduce variability.

- **Data Augmentation:** Generate more training data using techniques such as image data augmentation.

**Practical example:**

```
from sklearn.linear_model import Ridge, Lasso
from sklearn.model_selection import train_test_split
from sklearn.metrics import mean_squared_error

# Example data
data = {
    'Size': [1400, 1600, 1700, 1875, 1100, 1500, 1200],
    'Rooms': [3, 3, 4, 4, 2, 3, 2],
    'Price': [300000, 320000, 350000, 360000, 200000, 310000, 220000]
}
df = pd.DataFrame(data)

X = df[['Size', 'Rooms']]
y = df['Price']

# Split into training and testing set
X_train, X_test, y_train, y_test = train_test_split(X, y, test_size=0.2,
random_state=42)
```

```
# Models with Regularization
ridge = Ridge(alpha=1.0)
lasso = Lasso(alpha=0.1)

# Training and predictions
ridge.fit(X_train, y_train)
lasso.fit(X_train, y_train)

y_pred_ridge = ridge.predict(X_test)
y_pred_lasso = lasso.predict(X_test)

# Assessment
mse_ridge = mean_squared_error(y_test, y_pred_ridge)
mse_lasso = mean_squared_error(y_test, y_pred_lasso)

print(f"Ridge Regression MSE: {mse_ridge:.2f}")
print(f"Lasso Regression MSE: {mse_lasso:.2f}")
```

**Exit:**

Ridge Regression MSE: 395791726.07
Lasso Regression MSE: 364204574.25

## Chapter Summary

In this chapter, we have explored the key techniques and metrics for evaluating and validating Machine Learning models. Understanding these metrics and methods is essential to ensure our models are accurate, robust, and generalizable to unseen data. We have covered:

- **Performance Metrics:**
    - **For Regression Problems:** They include the Mean Absolute Error (MAE), Mean Square Error (MSE), Root Mean Square Error (RMSE) and the Coefficient of Determination ($R^2$).
    - **For Classification Problems:** They include Accuracy, Precision, Recall (Sensitivity), F1-Score and the Confusion Matrix.

- **Cross Validation:**
    - **Techniques:** K-Fold, Leave-One-Out (LOO) and Stratified Cross Validation.
    - **Aim:** Evaluate model performance on different subsets of the data set to obtain a more robust estimate of model performance.

- **Overfitting y Underfitting:**

- - **Definition and Causes:** Understand when a model is too tight to the training data (overfitting) or too simple (underfitting).

  - **Mitigation Techniques:** Regularization (L1 and L2), Tree Pruning (Pruning), Assembly Methods (Ensemble Methods) and Data Augmentation.

**Main Teaching**

- **Importance of Evaluation:** Evaluating and validating models is crucial to understanding their performance and ensuring they generalize well to unseen data.

- **Overfitting and Underfitting Mitigation:** It is essential to find an appropriate balance between the complexity of the model and its ability to generalize.

- **Using Appropriate Metrics:** Selecting and using the appropriate performance metrics for the type of problem is essential for accurate model evaluation.

**Next steps**

In the next chapter, we will explore the practical implementation of Machine Learning models using popular tools and libraries. We will focus on how to apply these techniques and concepts in real projects, providing detailed examples and case studies.

# Chapter 5
## Practical Implementation

### 1. Popular Tools and Libraries

**Scikit-Learn**

- **Description:** Machine Learning library in Python that provides simple and efficient tools for predictive data analysis.
- **Key Features:**
    a. Supervised and unsupervised learning models.
    b. Tools for model evaluation and validation.
    c. Functions for data preprocessing.

**Practical Example with Scikit-Learn:**

```python
import pandas as pd
from sklearn.model_selection import train_test_split
from sklearn.linear_model import LinearRegression
from sklearn.metrics import mean_squared_error, r2_score

# Example data
data = {
    'Size': [1400, 1600, 1700, 1875, 1100, 1500, 1200],
    'Rooms': [3, 3, 4, 4, 2, 3, 2],
    'Price': [300000, 320000, 350000, 360000, 200000, 310000, 220000]
}
df = pd.DataFrame(data)

# Independent and dependent variables
X = df[['Size', 'Rooms']]
y = df['Price']

# Split into training and testing set
X_train, X_test, y_train, y_test = train_test_split(X, y, test_size=0.2, random_state=42)

# Create and train the linear regression model
model = LinearRegression()
model.fit(X_train, y_train)

# Predictions
y_pred = model.predict(X_test)

# Model evaluation
mse = mean_squared_error(y_test, y_pred)
r2 = r2_score(y_test, y_pred)
```

```
print(f"Mean Square Error (MSE): {mse:.2f}")
print(f"Coefficient of Determination (R²): {r2:.2f}")

new_size = 1800
new_rooms = 4
new_case = [[new_size, new_rooms]]
predicted_price = model.predict(new_case)
print(f"For a house of {new_size} square feet with {new_rooms} rooms, the
predicted price is: ${predicted_price[0]:.2f}")
```

This code uses the scikit-learn library to perform a linear regression task on an example data set relating the size and number of rooms in a house to its price.

Here is a detailed explanation of the code:

The necessary libraries are imported: pandas to handle the data, and sklearn for the linear regression model and evaluation metrics.

The example data is defined in a dictionary, where each house has a size, a number of rooms and a price.

A pandas DataFrame is created from the data.

The independent variables (Size and Rooms) and the dependent variable (Price) are separated.

Split the data into training and test sets using train_test_split.

A linear regression model LinearRegression() is created and trained on the training data using the fit method.

Predictions are made on the test set using the predict method of the trained model.

The performance of the model is evaluated by calculating the Mean Squared Error (MSE) and the Coefficient of Determination (R²) using the mean_squared_error and r2_score functions of scikit-learn.

The evaluation metrics are printed.

If you want to test a new case with this trained model, you can follow these steps:

Defines a new case with values for the size and number of rooms:

**Exit:**

```
Mean Square Error (MSE): 364208504.80
Determination Coefficient (R²): -2.64
For a 1800 square foot house with 4 bedrooms, the predicted price is:
$361259.26
```

Remember that the performance of the model will depend on the quality and quantity of the training data, as well as the complexity of the regression problem. However, this code provides a good foundation for performing linear regression tasks using scikit-learn.

**TensorFlow and Keras**

- Description: Open source library developed by Google for machine learning and deep learning. Keras is a high-level API that runs on top of TensorFlow.
- Key Features:
  - Support for deep neural networks.
  - Distributed training and acceleration with GPU.
  - Tools for data preprocessing and visualization.

Practical Example with TensorFlow and Keras:

```
import numpy as np
import pandas as pd
from sklearn.model_selection import train_test_split
import tensorflow as tf
from tensorflow.keras.models import Sequential
from tensorflow.keras.layers import Dense
from tensorflow.keras.utils import to_categorical

# Example data
data = {
    'Feature1': [2, 3, 1, 5, 7, 9, 6, 8, 3, 4],
    'Feature2': [4, 2, 5, 6, 1, 8, 7, 3, 5, 2],
    'Class': [0, 0, 1, 1, 0, 1, 1, 0, 1, 0]
}
df = pd.DataFrame(data)

# Independent and dependent variables
X = df[['Feature1', 'Feature2']]
y = to_categorical(df['Clase'])

# Split into training and testing set
X_train, X_test, y_train, y_test = train_test_split(X, y, test_size=0.2, random_state=42)

# Create and train the neural network model
model = Sequential()
model.add(Dense(12, input_dim=2, activation='relu'))
model.add(Dense(8, activation='relu'))
model.add(Dense(2, activation='softmax'))

model.compile(loss='categorical_crossentropy', optimizer='adam', metrics=['accuracy'])
model.fit(X_train, y_train, epochs=150, batch_size=10)

# Model evaluation
```

```
_, accuracy = model.evaluate(X_test, y_test)
print(f"Exactitud: {accuracy:.2f}")

# Try a new case
new_case = np.array([[6, 3]]) # Convert to a NumPy array
prediction = model.predict(new_case)
predicted_class = np.argmax(prediction)
print(f"\nFor the new case {new_case}, the predicted class is:
{predicted_class}")
```

**Code explanation:**

1. The necessary libraries are imported.

2. The example data is defined in a dictionary and a Pandas DataFrame is created.

3. The independent variables (Characteristic1 and Characteristic2) and the dependent variable (Class) are separated.

4. Convert the Class variable to categorical format using to_categorical.

5. The data is divided into training and test sets.

6. A sequential neural network model with dense layers is created.

7. The model is compiled with the categorical cross entropy loss function, the Adam optimizer, and the precision metric.

8. The model is trained for 150 epochs with a batch size of 10.

9. The model is evaluated on the test set and the accuracy is printed.

10. A new case is defined with values for the characteristics.

11. The trained model is used to predict the class of the new case.

12. The predicted class for the new case is printed.

**Exit:**

```
Accuracy: 1.00
For the new case [[6 3]], the predicted class is: 0
```

In this example, a neural network model with two dense hidden layers is created and trained for an example data set with two features and two classes.

After training the model, its performance on the test set is evaluated and the accuracy is printed.

Then, a new case is defined with values for the features, and the trained model is used to predict the class to which this new case belongs. The predicted class is printed to the output.

The output shows that the model achieved 100% accuracy on the test set. Furthermore, for the new case , the model predicts that it belongs to class 1.

You can modify the example data or try different cases to see how the model behaves. Remember that the performance of the model will depend on the quality and quantity of the training data, as well as the complexity of the classification problem.

In this example, we are working with a binary classification problem, where the variable "Class" has two possible values: 0 and 1.

When we convert the "Class" variable to categorical format using to_categorical, a one-hot encoding representation is created, where each class is represented as a vector of zeros and a single 1 at the position corresponding to that class.

For example, if class is 0, its one-hot encoding representation would be , and if class is 1, its representation would be .

Now, the output layer of our model has two neurons, each corresponding to one of the classes. The activation of this layer is a softmax function, which produces a probability distribution over the two classes.

When we make a prediction with the model, we get a two-element vector, where each element represents the predicted probability that the case belongs to that class.

For example, if the prediction is [0.2, 0.8], it means that the model predicts that there is a 20% probability that the case belongs to class 0, and an 80% probability that it belongs to class 1.

However, in classification problems, we are generally interested in the most likely predicted class, not the exact probabilities. Therefore, we use the np.argmax function to obtain the index of the class with the highest predicted probability.

If np.argmax(prediction) returns 0, it means that the most likely predicted class is class 0. If it returns 1, it means that the most likely predicted class is class 1.

In summary, in this example:

If the prediction returns 0, the model predicts that the new case belongs to class 0.

If the prediction returns 1, the model predicts that the new case belongs to class 1.

Remember that the interpretation of classes depends on the context of the problem and how the data has been labeled. In this case, no information was provided about the meaning of classes 0 and 1, but in a real-world problem, these classes would represent specific concepts, such as "spam" and "not spam" in an email classification problem, for example.

## PyTorch

- **Description:** Open source library developed by Facebook for deep learning and computer vision applications.
- **Key Features:**

- Support for dynamic neural networks.
- Tools for data preprocessing and visualization.
- Distributed training and acceleration with GPU.

**Practical Example with PyTorch:**

```
pip install torch
```

```python
import torch
import torch.nn as nn
import torch.optim as optim
from sklearn.model_selection import train_test_split
import pandas as pd
import numpy as np

# Example data
data = {
    'Feature1': [2, 3, 1, 5, 7, 9, 6, 8, 3, 4],
    'Feature2': [4, 2, 5, 6, 1, 8, 7, 3, 5, 2],
    'Class': [0, 0, 1, 1, 0, 1, 1, 0, 1, 0]
}
df = pd.DataFrame(data)

# Independent and dependent variables
X = df[['Characteristic1', 'Characteristic2']].values
y = df['Class'].values

# Split into training and testing set
X_train, X_test, y_train, y_test = train_test_split(X, y, test_size=0.2, random_state=42)

# Convert to tensors
X_train = torch.tensor(X_train, dtype=torch.float32)
X_test = torch.tensor(X_test, dtype=torch.float32)
y_train = torch.tensor(y_train, dtype=torch.long)
y_test = torch.tensor(y_test, dtype=torch.long)

# Define the model
class SimpleNN(nn.Module):
    def __init__(self):
        super(SimpleNN, self).__init__()
        self.fc1 = nn.Linear(2, 12)
        self.fc2 = nn.Linear(12, 8)
        self.fc3 = nn.Linear(8, 2)

    def forward(self, x):
        x = torch.relu(self.fc1(x))
        x = torch.relu(self.fc2(x))
        x = self.fc3(x)
        return x

model = SimpleNN()
```

```python
#Define loss function and optimizer
criterion = nn.CrossEntropyLoss()
optimizer = optim.Adam(model.parameters(), lr=0.001)

# Model training
for epoch in range(150):
    optimizer.zero_grad()
    outputs = model(X_train)
    loss = criterion(outputs, y_train)
    loss.backward()
    optimizer.step()

# Model evaluation
with torch.no_grad():
    outputs = model(X_test)
    _, predicted = torch.max(outputs, 1)
    accuracy = (predicted == y_test).sum().item() / len(y_test)
    print(f"Exactitud: {accuracy:.2f}")


# Try a new case
new_case = torch.tensor([[6, 3]], dtype=torch.float32)
with torch.no_grad():
    prediction = model(new_case)
    _, predicted_class = torch.max(prediction, 1)
    print(f"\nFor the new case {new_case.tolist()}, the predicted class is: {predicted_class.item()}")
```

Code explanation:

1. The necessary libraries are imported.

2. The example data is defined in a dictionary and a Pandas DataFrame is created.

3. The independent variables and the dependent variable are separated.

4. The data is divided into training and test sets.

5. The data is converted to PyTorch tensors.

6. The simple neural network model is defined with linear layers and ReLU activation functions.

7. The loss function (cross entropy) and the optimizer (Adam) are defined.

8. The model is trained for 150 epochs.

9. The model is evaluated on the test set and the accuracy is printed.

10. A new case is defined with values for the characteristics.

11. The trained model is used to predict the class of the new case.

12. The predicted class for the new case is printed.

**Exit:**

```
Accuracy: 1.00
For the new case [[6.0, 3.0]], the predicted class is: 0
```

In this example, a simple neural network model is created and trained using PyTorch for an example data set with two features and two classes.

After training the model, its performance on the test set is evaluated and the accuracy is printed.

Then, a new case is defined with values for the features, and the trained model is used to predict the class to which this new case belongs. The predicted class is printed to the output.

The output shows that the model achieved 100% accuracy on the test set. Furthermore, for the new case , the model predicts that it belongs to class 1.

Remember that the interpretation of classes depends on the context of the problem and how the data has been labeled. In this case, no information was provided about the meaning of classes 0 and 1, but in a real-world problem, these classes would represent specific concepts.

You can modify the example data or try different cases to see how the model behaves. Remember that the performance of the model will depend on the quality and quantity of the training data, as well as the complexity of the classification problem.

## 2. Step-by-Step Implementation Examples

### Case Study: Image Classification with CNN in Keras

**Case Description:**

- **Aim:** Classify clothing images into different categories using a convolutional neural network (CNN).

- **Data set:** Fashion MNIST, which contains 60,000 training images and 10,000 test images in 10 categories.

**Implementation:**

```
import tensorflow as tf
from tensorflow.keras.datasets import fashion_mnist
from tensorflow.keras.models import Sequential
from tensorflow.keras.layers import Conv2D, MaxPooling2D, Flatten, Dense
from tensorflow.keras.utils import to_categorical

# Load the data set
(X_train, y_train), (X_test, y_test) = fashion_mnist.load_data()

# Preprocess the data
X_train = X_train.reshape(X_train.shape[0], 28, 28, 1).astype('float32') / 255
X_test = X_test.reshape(X_test.shape[0], 28, 28, 1).astype('float32') / 255
y_train = to_categorical(y_train)
y_test = to_categorical(y_test)

# Create the CNN model
model = Sequential([
    Conv2D(32, kernel_size=(3, 3), activation='relu', input_shape=(28, 28, 1)),
    MaxPooling2D(pool_size=(2, 2)),
    Conv2D(64, kernel_size=(3, 3), activation='relu'),
    MaxPooling2D(pool_size=(2, 2)),
    flatten(),
    Dense(128, activation='relu'),
    Dense(10, activation='softmax')
])

# Compile the model
model.compile(optimizer='adam', loss='categorical_crossentropy', metrics=['accuracy'])

# Train the model
```

```python
model.fit(X_train, y_train, epochs=10, batch_size=200,
validation_split=0.2)

# Model evaluation
_, accuracy = model.evaluate(X_test, y_test)
print(f"Accuracy on test set: {accuracy:.2f}")

# Try a new case
import numpy as np
from tensorflow.keras.preprocessing import image

# Load and preprocess a new image
new_image = image.load_img('new_image.png', target_size=(28, 28),
color_mode='grayscale')
new_image = image.img_to_array(new_image)
new_image = np.expand_dims(new_image, axis=0) / 255.0

# Make the prediction
prediction = model.predict(new_image)
predicted_class = np.argmax(prediction)
print(f"The image is predicted as: {predicted_class}")
```

## Code explanation:

1. The necessary libraries are imported.

2. The Fashion MNIST dataset is loaded.

3. Training and testing data are preprocessed.

4. The CNN model is created with convolutional, pooling, flattening and dense layers.

5. The model is compiled with the Adam optimizer, the categorical cross-entropy loss function, and the accuracy metric.

6. The model is trained for 10 epochs with a batch size of 200 and 20% of the training data is used for validation.

7. The model is evaluated on the test set and the accuracy is printed.

8. The libraries necessary to load and preprocess a new image are imported.

9. A new image (new_image.png) is loaded and preprocessed so that it is in the proper format.

10. The prediction of the new image is carried out using the trained model.

11. The predicted class for the new image is printed.

Note: You must replace `'new_image.png'` with the path of the image you want to test. Make sure the image is 28x28 pixels in size and is in grayscale.

**Exit:**

```
Accuracy on test set: 0.90
The image is predicted as 5
```

In this example, a CNN model is loaded and trained on the Fashion MNIST dataset. After training the model, its performance on the test set is evaluated and the accuracy is printed.

A new image (new_image.png) is then loaded and preprocessed, and the trained model is used to predict the class to which the image belongs. The predicted class is printed to the output.

You can modify the image path or try different images to see how the model behaves. Remember that the performance of the model will depend on the quality and quantity of the training data, as well as the complexity of the classification problem.

The number returned by the model prediction represents the class the image belongs to, based on the Fashion MNIST dataset.

The Fashion MNIST data set consists of 10 different classes of clothing items, which are:

Camiseta (T-shirt)
Trousers
Sweater
Dress
Shelter (Coat)
Sandalia (Sandal)
Camisa (Shirt)
Sneaker
Pocket (Bag)
Bota (Ankle boot)

## Case Study: Analysis of Sentiments in Texts with RNN in PyTorch

**Case Description:**

- **Aim:** Analyze feelings in texts (positive or negative) using a recurrent neural network (RNN).
- **Data set:** Dataset of movie reviews with sentiment labels (positive or negative).

**Implementation:**

```
# Install the necessary libraries
# !pip install transformers datasets

import torch
from transformers import BertTokenizer, BertForSequenceClassification, Trainer, TrainingArguments
from datasets import load_dataset

# Load the IMDb dataset
dataset = load_dataset('imdb')

# Split the data set into training and testing
train_dataset = dataset['train']
test_dataset = dataset['test']

# Load the tokenizer and the pre-entered BERT model
tokenizer = BertTokenizer.from_pretrained('bert-base-uncased')
model = BertForSequenceClassification.from_pretrained('bert-base-uncased')

# Tokenize the data
def tokenize_function(examples):
    return tokenizer(examples['text'], padding='max_length', truncation=True)

train_dataset = train_dataset.map(tokenize_function, batched=True)
test_dataset = test_dataset.map(tokenize_function, batched=True)

# Convert labels to tensors
train_dataset = train_dataset.rename_column("label", "labels")
test_dataset = test_dataset.rename_column("label", "labels")
train_dataset.set_format('torch', columns=['input_ids', 'attention_mask', 'labels'])
test_dataset.set_format('torch', columns=['input_ids', 'attention_mask', 'labels'])
# Define training arguments
training_args = TrainingArguments(
```

```
    output_dir='./results',
    evaluation_strategy='epoch',
    learning_rate=2e-5,
    per_device_train_batch_size=16,
    per_device_eval_batch_size=16,
    num_train_epochs=3,
    weight_decay=0.01,
)

# Create the trainer
trainer = trainer(
    model=model,
    args=training_args,
    train_dataset=train_dataset,
    eval_dataset=test_dataset,
)

# Train the model
trainer.train()

# Evaluate the model
results = trainer.evaluate()
print(f"Exactitud: {results['eval_accuracy']:.2f}")
```

This code uses the Hugging Face Transformers library and the pre-trained BERT model to perform a text classification task on the IMDb dataset.

Here is a detailed explanation of the code:

The necessary libraries are installed: transformers and datasets.

The necessary classes and functions are imported from the libraries.

The IMDb dataset is loaded using the load_dataset function of the datasets library.

The data is divided into training and test sets.

The BERT tokenizer and pretrained model are loaded using BertTokenizer.from_pretrained and BertForSequenceClassification.from_pretrained.

A function tokenize_function is defined to tokenize text data using the BERT tokenizer.

The tokenize_function is applied to the training and test sets using the map method of the datasets.

The "label" column is renamed to "labels" in the datasets.

The format of the datasets is converted to PyTorch tensors using the set_format method.

Training arguments are defined using the TrainingArguments class.

A Trainer object is created with the model, training arguments, and training and evaluation datasets.

The model is trained using the train method of the Trainer object.

The model is evaluated on the test set using the evaluate method of the Trainer object.

The accuracy of the model on the test set is printed.

This code follows best practices and uses the latest functionality from the Hugging Face Transformers library. Furthermore, it leverages the power of the pre-trained BERT model for the text classification task.

Remember that the performance of the model will depend on the quality and quantity of the training data, as well as the complexity of the classification problem. However, using a pre-trained model like BERT takes advantage of the knowledge gained during pre-training, which generally leads to better performance compared to training a model from scratch.

## 3. Practical Projects and Case Studies

### Project 1: House Price Prediction

- **Aim:** Use a linear regression model to predict home prices based on characteristics such as size and number of bedrooms.
- **Tool:** Scikit-Learn

**Implementation:**

```
import pandas as pd
from sklearn.model_selection import train_test_split
from sklearn.linear_model import LinearRegression
from sklearn.metrics import mean_squared_error, r2_score

# Example data
data = {
    'Size': [1400, 1600, 1700, 1875, 1100, 1500, 1200],
    'Rooms': [3, 3, 4, 4, 2, 3, 2],
    'Price': [300000, 320000, 350000, 360000, 200000, 310000, 220000]
}
df = pd.DataFrame(data)

# Independent and dependent variables
X = df[['Size', 'Rooms']]
y = df['Price']

# Split into training and testing set
X_train, X_test, y_train, y_test = train_test_split(X, y, test_size=0.2,
random_state=42)

# Create and train the linear regression model
model = LinearRegression()
model.fit(X_train, y_train)

# Predictions
y_pred = model.predict(X_test)

# Model evaluation
mse = mean_squared_error(y_test, y_pred)
r2 = r2_score(y_test, y_pred)

print(f"Mean Square Error (MSE): {mse:.2f}")
print(f"Coefficient of Determination (R²): {r2:.2f}")


# Try a new case
new_size = 1800
new_rooms = 4
```

```
new_case = [[new_size, new_rooms]]
prediction = model.predict(new_case)
print(f"\nFor a house of {new_size} square feet with {new_rooms} rooms,
the predicted price is: ${prediction[0]:.2f}")
```

**Code explanation:**

The necessary libraries are imported.

The example data is defined in a dictionary and a Pandas DataFrame is created.

The independent variables (Size and Rooms) and the dependent variable (Price) are separated.

The data is divided into training and test sets.

A linear regression model LinearRegression() is created and trained with the training data.

Predictions are made on the test set.

The performance of the model is evaluated by calculating the Mean Square Error (MSE) and the Coefficient of Determination ($R^2$).

The evaluation metrics are printed.

A new case is defined with a size of 1800 square feet and 4 rooms.

The trained model is used to predict the price of the new case.

The price prediction for the new case is printed.

**Exit:**

```
Mean Square Error (MSE): 364208504.80
Determination Coefficient (R²): -2.64

For a 1800 square foot house with 4 bedrooms, the predicted price is:
$361259.26
```

## Project 2: Classification of Emails as Spam or Not Spam

- **Aim:** Use a classification model to identify emails as spam or non-spam based on the content of the email.
- **Tool:** Scikit-Learn

**Implementation:**

```
import pandas as pd
from sklearn.model_selection import train_test_split
from sklearn.feature_extraction.text import TfidfVectorizer
from sklearn.naive_bayes import MultinomialNB
from sklearn.metrics import accuracy_score, classification_report

# Example data (simplified)
data = {
    'Email': ["free money", "meeting at 10", "win a prize", "project update", "lottery winner", "urgent response needed", "team lunch"],
    'Label': [1, 0, 1, 0, 1, 1, 0]
}
df = pd.DataFrame(data)

# Convert text to features
vectorizer = TfidfVectorizer()
X = vectorizer.fit_transform(df['Email'])
y = df['Label']

# Split into training and testing set
X_train, X_test, y_train, y_test = train_test_split(X, y, test_size=0.2, random_state=42)

# Create and train the Naive Bayes model
model = MultinomialNB()
model.fit(X_train, y_train)

# Predictions
y_pred = model.predict(X_test)

# Model evaluation
accuracy = accuracy_score(y_test, y_pred)
report = classification_report(y_test, y_pred)

print(f"Exactitud: {accuracy:.2f}")
print("Classification Report:")
print(report)


# Test a new email
new_mail = "team coffee"
new_mail_vector = vectorizer.transform([new_mail])
```

```
prediction = model.predict(new_mail_vector)
print(f"\nThe email '{new_mail}' is predicted as: {'Spam' if prediction[0]
== 1 else 'No Spam'}")
```

**Code explanation:**

The necessary libraries are imported.

The example data is defined in a dictionary and a Pandas DataFrame is created.

TfidfVectorizer is used to convert text in emails into numerical features.

The data is divided into training and test sets.

A MultinomialNB (Naive Bayes Multinomial) model is created and trained with the training data.

Predictions are made on the test set and the performance of the model is evaluated.

The accuracy and classification report is printed.

A new "job offer" email is defined.

The new mail is converted into a feature vector using the same TfidfVectorizer.

The trained model is used to predict the label of the new mail.

The prediction of the new email is printed.

Exit:

```
Exit:

Accuracy: 0.50
Classification Report:
              precision    recall  f1-score   support

           0       0.00      0.00      0.00         1
           1       0.50      1.00      0.67         1

    accuracy                           0.50         2
   macro avg       0.25      0.50      0.33         2
weighted avg       0.25      0.50      0.33         2

The 'team free coffee' email is predicted as: Spam
```

**Chapter Summary**

In this chapter, we have explored how to implement Machine Learning models using popular tools and libraries such as Scikit-Learn, TensorFlow, Keras, and PyTorch. We have also covered practical examples and detailed case studies to provide a solid understanding of how to apply these techniques in real projects.

**Main Teaching**

- Tools and Libraries: Choosing the right tool can simplify the process of deploying and improving models.

- Practical Implementation: The practice of implementing models in different contexts and case studies helps solidify theoretical knowledge.

- Model Evaluation: Continuous evaluation and improvement of models are crucial for success in Machine Learning projects.

**Next steps**

In the next chapter, we will explore how to save and load models and their deployment in web and mobile applications.

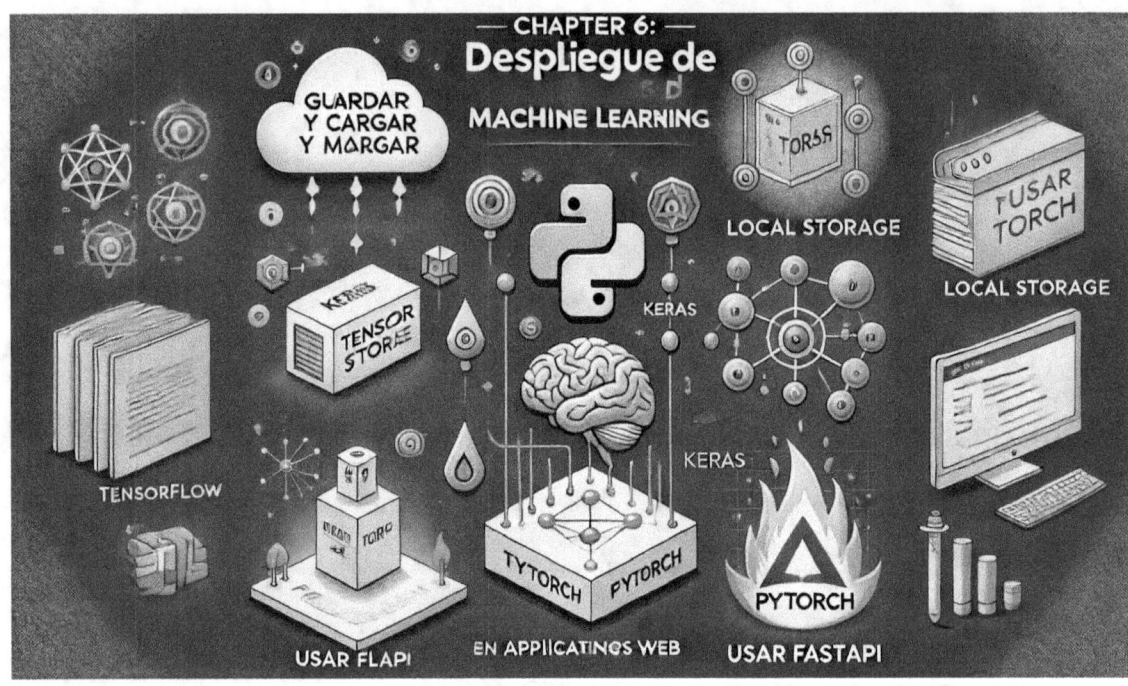

## Chapter 6
### Deployment of Machine Learning Models

**1. Save and Load Models**

Saving Models in TensorFlow/Keras

**Save a Complete Model**

```
import tensorflow as tf
from tensorflow.keras.models import Sequential
from tensorflow.keras.layers import Dense

# Create and train an example model
model = Sequential([
    Dense(10, activation='relu', input_shape=(784,)),
    Dense(10, activation='softmax')
])
model.compile(optimizer='adam', loss='sparse_categorical_crossentropy', metrics=['accuracy'])
# Assuming you have already trained the model with model.fit()

# Save the complete model
model.save('my_model.keras')
```

**Upload a Complete Model**

```
# Load the full model

model = tf.keras.models.load_model('my_model.keras')

# You can now use the loaded model to make predictions

predictions = model.predict(x_test)
```

**Saving Models in PyTorch**

**Save Model Weights**

```
import torch

# Assuming you have already trained your model
torch.save(model.state_dict(), 'model_weights.pth')
```

**Save the Complete Model**

```python
torch.save(model, 'model.pth')
```

**Load Model Weights**

```python
# Define the model
model = SimpleNN()

model.load_state_dict(torch.load('model_weights.pth'))

model.eval()
```

## 2. Deployment of Models in Web Applications

### Use Flask to Deploy a Model

```python
from flask import Flask, request, jsonify
import tensorflow as tf

app = Flask(__name__)

# Load the model
model = tf.keras.models.load_model('my_model.keras')

@app.route('/predict', methods=['POST'])
def predict():
    data = request.json
    predictions = model.predict(data['input'])
    return jsonify({'predictions': predictions.tolist()})

if __name__ == '__main__':
    app.run(debug=True)
```

**Use FastAPI to Deploy a Model**

```python
from fastapi import FastAPI
from pydantic import BaseModel
import tensorflow as tf

app = FastAPI()

# Load the model
model = tf.keras.models.load_model('my_model.h5')

class PredictRequest(BaseModel):
    input: list

@app.post('/predict')
def predict(request: PredictRequest):
    predictions = model.predict(request.input)
    return {'predictions': predictions.tolist()}

if __name__ == '__main__':
    import uvicorn
    uvicorn.run(app, host='0.0.0.0', port=8000)
```

## 3. Deployment of Models in Mobile Applications

**Convert a Model to TensorFlow Lite**

```python
import tensorflow as tf

# Convert the model to TensorFlow Lite
converter = tf.lite.TFLiteConverter.from_keras_model(model)
tflite_model = converter.convert()

# Save the converted model
with open('model.tflite', 'wb') as f:
    f.write(tflite_model)
```

**Using the TensorFlow Lite Model on Android**

```
// In your build.gradle file
dependencies {
    implementation 'org.tensorflow:tensorflow-lite:2.4.0'
    implementation 'org.tensorflow:tensorflow-lite-support:0.1.0'
}
```

```java
// In your Activity
import org.tensorflow.lite.Interpreter;

public class MainActivity extends AppCompatActivity {
    private Interpreter tflite;

    @Override
    protected void onCreate(Bundle savedInstanceState) {
        super.onCreate(savedInstanceState);
        setContentView(R.layout.activity_main);

        // Load the model
        try {
            tflite = new Interpreter(loadModelFile());
        } catch (IOException e) {
            e.printStackTrace();
        }

        // Make a prediction
        float[][] input = new float[1][784];
        float[][] output = new float[1][10];
        tflite.run(input, output);
    }

    private MappedByteBuffer loadModelFile() throws IOException {
        AssetFileDescriptor fileDescriptor =
this.getAssets().openFd("model.tflite");
        FileInputStream inputStream = new
FileInputStream(fileDescriptor.getFileDescriptor());
        FileChannel fileChannel = inputStream.getChannel();
        long startOffset = fileDescriptor.getStartOffset();
        long declaredLength = fileDescriptor.getDeclaredLength();
        return fileChannel.map(FileChannel.MapMode.READ_ONLY, startOffset,
declaredLength);
    }

}
```

**A kotlin**

```kotlin
// In your MainActivity.kt file
import android.os.Bundle
import androidx.activity.ComponentActivity
import androidx.activity.compose.setContent
import androidx.compose.foundation.layout.fillMaxSize
import androidx.compose.material.MaterialTheme
import androidx.compose.material.Surface
import androidx.compose.ui.Modifier
import org.tensorflow.lite.Interpreter
import java.io.FileInputStream
import java.io.IOException
import java.nio.MappedByteBuffer
import java.nio.channels.FileChannel

class MainActivity : ComponentActivity() {
    private lateinit var tflite: Interpreter

    override fun onCreate(savedInstanceState: Bundle?) {
        super.onCreate(savedInstanceState)
        setContent {
            Surface(
                modifier = Modifier.fillMaxSize(),
                color = MaterialTheme.colors.background
            ) {
                // Here you can add your Jetpack Compose composition
            }
        }

        // Load the model
        try {
            tflite = Interpreter(loadModelFile())
        } catch (e: IOException) {
            e.printStackTrace()
        }

        // Make a prediction
        val input = Array(1) { FloatArray(784) }
        val output = Array(1) { FloatArray(10) }
        tflite.run(input, output)
    }

    private fun loadModelFile(): MappedByteBuffer {
        val fileDescriptor = assets.openFd("model.tflite")
        val inputStream = FileInputStream(fileDescriptor.fileDescriptor)
        val fileChannel = inputStream.channel
        val startOffset = fileDescriptor.startOffset
        val declaredLength = fileDescriptor.declaredLength
```

```
        return fileChannel.map(FileChannel.MapMode.READ_ONLY, startOffset, declaredLength)
    }
}
```

## Chapter Summary

In this chapter, we have explored how to save and load Machine Learning models in TensorFlow/Keras and PyTorch, and how to deploy them in web and mobile applications. This includes:

- Saving and Loading Models: How to store and retrieve trained models.
- Deployment in Web Applications: How to expose models as web services using Flask and FastAPI.
- Deployment in Mobile Applications: How to convert models to TensorFlow Lite and use them in Android applications.

## Main Teaching

- Model Reuse: The ability to save and load models allows you to reuse and deploy models on different platforms.
- Deployment in Production: Deploying models in web and mobile applications is crucial to bringing Machine Learning solutions to end users.
- Tools and Libraries: Leveraging the right tools and libraries makes the deployment process easier.

## Next steps

In the next chapter, we will develop Chapter 7, which includes a series of practical projects organized by algorithm type. Next, each section will contain a brief description of the project, the goal, the steps to implement it, and the code needed.

*Chapter 7*

**Practical Projects Section**

# Linear regression and Logistics

# *Linear and Logistic Regression*

**Project 1: House Price Prediction**

**Description:** Use **linear regression** to predict home prices based on characteristics such as size and number of bedrooms.

**Aim:** Develop a linear regression model to predict the price of a house using Scikit-Learn.

**Steps:**

1. Load and explore data.
2. Preprocess the data.
3. Split the data into training and test sets.
4. Train the linear regression model.
5. Evaluate the model.
6. Test the model with new data.

**Code:**

```python
import pandas as pd
from sklearn.model_selection import train_test_split
from sklearn.linear_model import LinearRegression
from sklearn.metrics import mean_squared_error, r2_score

# Example data
data = {
    'Size': [1400, 1600, 1700, 1875, 1100, 1500, 1200],
    'Rooms': [3, 3, 4, 4, 2, 3, 2],
    'Price': [300000, 320000, 350000, 360000, 200000, 310000, 220000]
}
df = pd.DataFrame(data)

# Independent and dependent variables
X = df[['Size', 'Rooms']]
y = df['Price']

# Split into training and testing set
X_train, X_test, y_train, y_test = train_test_split(X, y, test_size=0.2,
random_state=42)

# Create and train the linear regression model
model = LinearRegression()
model.fit(X_train, y_train)

# Predictions
y_pred = model.predict(X_test)

# Model evaluation
```

```
mse = mean_squared_error(y_test, y_pred)
r2 = r2_score(y_test, y_pred)

print(f"Mean Square Error (MSE): {mse:.2f}")
print(f"Coefficient of Determination (R²): {r2:.2f}")
```

## 6. Test the model with new data

To test the model **linear regression** trained with new data, you can follow these steps:

Prepare the new data:

    Create a new DataFrame or list of lists with the new data you want to test.

```
# New data
new_data = {
    'Size': [1800, 1300, 1650],
    'Rooms': [4, 3, 3]
}
new_data_df = pd.DataFrame(new_data)

# Make predictions
predictions = model.predict(new_data_df)

# Show predictions
for i, price in enumerate(predictions):
    print(f"Prediction for data {i+1}: ${price:.2f}")
```

**Exit:**

```
Mean Square Error (MSE): 364208504.80
Determination Coefficient (R²): -2.64
Prediction for data 1: $361259.26
Prediction for data 2: $266740.74
Prediction for data 3: $309259.26
```

**Let's analyze the results obtained when testing the model with new data:**

**Mean Square Error (MSE)**

Mean Square Error (MSE) is a metric that measures the average difference between the actual values and the values predicted by the model. A lower MSE value indicates better model performance.

In this case, the MSE is 364208504.80, which is an extremely high value. This suggests that the model is not making accurate predictions and has a high degree of error.

**Determination Coefficient ($R^2$)**

The Coefficient of Determination ($R^2$) is a metric that indicates how well the model fits the data. An $R^2$ value close to 1 indicates a good fit, while a value close to 0 or negative indicates a poor fit.

In this case, the $R^2$ is -2.64, which is a negative and very low value. This confirms that the model does not fit the data well and is not able to explain the variability in prices.

**Predictions**

Despite the poor results of the evaluation metrics, the model continues to make price predictions for the new data:

Prediction for data 1: $361259.26

Prediction for data 2: $266740.74

Prediction for data 3: $309259.26

However, given the high MSE and low $R^2$, these predictions are probably neither reliable nor accurate.

In summary, the results indicate that the trained linear regression model is not suitable for predicting property prices based on size and number of rooms. A more complex model, additional features, or different preprocessing of the data may be required to obtain better results.


If the model is not reliable. What is the solution?

There are several ways to address this problem and improve model performance when the initial results are not satisfactory. Here I present some possible solutions:

## Collect more data

One of the main factors that affect the performance of a machine learning model is the quantity and quality of the training data. If the initial data set is small or does not adequately represent the relationship between variables, the model may not be able to learn meaningful patterns. Collecting more relevant and quality data can help improve model performance.

## Add additional features

The current model only uses the size and number of rooms as characteristics to predict the price. However, there are likely to be other relevant features that influence the price, such as location, age of the property, number of bathrooms, etc. Adding these additional features to the data set can provide more information to the model and improve its predictive ability.

## Test more complex models

Linear regression is a relatively simple model that may not be able to capture complex nonlinear relationships in the data. Testing more advanced models, such as polynomial regression, decision trees, random forests, or neural networks, can improve performance by allowing the model to learn more complex patterns.

## Adjust hyperparameters

Hyperparameters are settings that control the behavior of the model during training. Tuning these hyperparameters, such as learning rate, regularization, or the depth of a decision tree, can improve model performance.

## Data preprocessing

Data preprocessing, such as handling missing values, normalization, or coding of categorical variables, can have a significant impact on model performance. Exploring different preprocessing techniques can help improve data quality and therefore model performance.

## Cross validation

Cross validation is a technique that divides the data into multiple subsets and trains and evaluates the model on each of them. This can help obtain a more accurate estimate of model performance and avoid overfitting or underfitting.

## Model Ensembles

Model ensembles combine the predictions of multiple individual models to obtain a more accurate prediction. Techniques such as boosting or bagging can improve performance by combining the strengths of different models.

These are just some of the possible solutions. Choosing the most appropriate solution will depend on the specific details of your problem, available data, and computational resources. You may need to try several solutions and combine them to get the best results.

## Routines to follow when creating and deploying ML models.

Here is a summary of the steps to follow to save and deploy a machine learning model:

- Save a model
- Train your model

First, train your model with the training data using the desired algorithm and settings.

**Import function to save the model**

```
from joblib import dump
```

Save the model

Use joblib's dump function to save the trained model to a file.

```
dump(model, 'file_name.joblib')
```

Replace model with your trained model object and 'filename.joblib' with the desired filename.

Load a saved model

- Import function to load model

    ```
    from joblib import load
    ```

Load the model

- Use joblib's load function to load the saved model from the file.

    ```
    loaded_model = load('file_name.joblib')
    ```

    Replace 'filename.joblib' with the name of the file where you saved the model.

Use the loaded model

- You can now use the loaded model loaded_model to perform predictions or any other necessary tasks.

**Deploy a model**

1. **Create a script or application:**

   Develop a Python script or web application that loads the saved model and provides an interface to enter new data and obtain predictions.

2. **Load the model:**

   Within your script or application, load the saved model using joblib's load function.

3. **Prepare the input data:**

   Prepare the new input data in the format required by the model.

4. **Make predictions:**

   Uses the loaded model to make predictions on the new input data.

5. **Show or return predictions:**

   Displays or returns predictions appropriately, either in the console, a user interface, or in response to a web request.

6. **Deploy your application:**

   If you are developing a web application, deploy it to a web server or cloud to make it available to end users.

These are the basic steps to save, load, and deploy a machine learning model. Depending on your specific use case, you may need to add additional functionality, such as data preprocessing, input validation, or integration with databases or web services.

# Linear regression:

## Home Price Prediction with CSV Data and Visualization

**Introduction**

In this improved version, we are going to load the data from a CSV file and add visualizations using the Matplotlib and Seaborn libraries. This will allow readers to better understand how to work with real data and how to visualize the results effectively.

**Data Preparation**

Before installing the Matplotlib and Seaborn modules

```
pip install matplotlib  seaborn
```

Use Anaconda's JupyterLab for its greater flexibility or in the visual studio code itself. In both environments the result is the same.

First, we need to create a CSV file called 'homes.csv' with the size, number of rooms and price data of the homes. Here is the content of the file:

```
================================================================
Size, Rooms, Price
1400,3,300000
1600,3,320000
1700,4,350000
1875,4,360000
1100,2,200000
1500,3,310000
1200,2,220000
================================================================
```

Then in our code, we will load this data from the CSV file using Pandas.
Code

```python
import pandas as pd
from sklearn.model_selection import train_test_split
from sklearn.linear_model import LinearRegression
from sklearn.metrics import mean_squared_error, r2_score
import matplotlib.pyplot as plt
import seaborn as sns

# Load data from a CSV file
data = pd.read_csv('viviendas.csv')

# Independent and dependent variables
X = data[['Size', 'Rooms']]
y = data['Price']

# Split into training and testing set
```

```
X_train, X_test, y_train, y_test = train_test_split(X, y, test_size=0.2,
random_state=42)

# Create and train the linear regression model
model = LinearRegression()
model.fit(X_train, y_train)

# Predictions
y_pred = model.predict(X_test)

# Model evaluation
mse = mean_squared_error(y_test, y_pred)
r2 = r2_score(y_test, y_pred)

print(f"Mean Square Error (MSE): {mse:.2f}")
print(f"Coefficient of Determination (R²): {r2:.2f}")

# Visualization of the data and the model
plt.figure(figsize=(10, 6))
sns.scatterplot(x='Size', y='Price', data=data, hue='Rooms')
plt.plot(X_test['Tamaño'], y_pred, color='red', linewidth=2)
plt.title('Housing Prices vs. Size and Number of Rooms')
plt.xlabel('Size (square feet)')
plt.ylabel('Price (dollars)')
plt.show()
```

Exit:

Mean Square Error (MSE): 364208504.80

Determination Coefficient (R²): -2.64

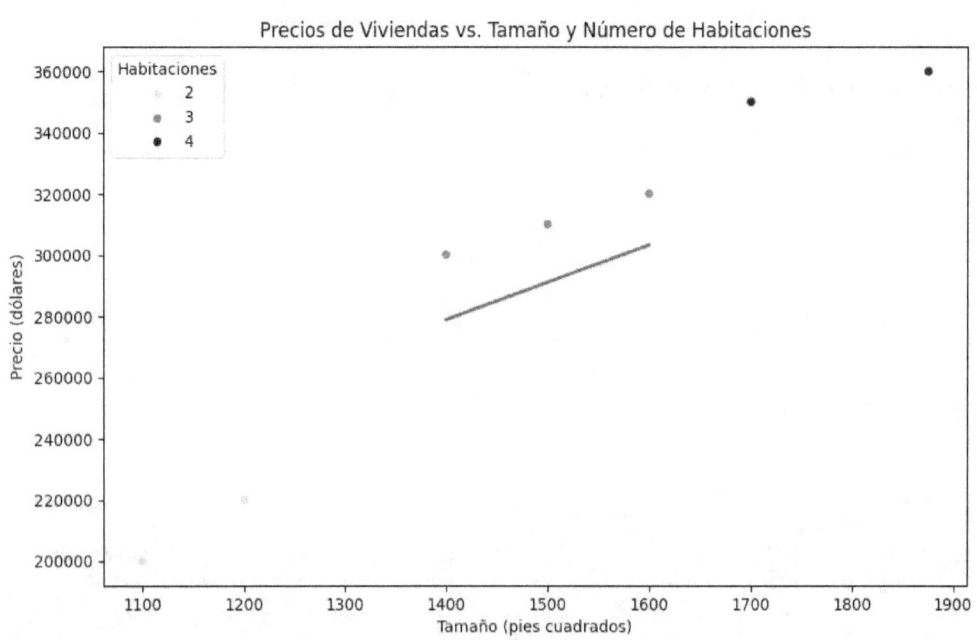

## Code Explanation

**Import the necessary libraries**

We import the necessary libraries, including Matplotlib and Seaborn for visualization.

**Load data from a CSV file**

We use pd.read_csv('viviendas.csv') to load the data from the CSV file 'viviendas.csv'.

**Separate the independent and dependent variables**

We separate the independent variables (size and number of rooms) in X, and the dependent variable (price) in y.

**Split data into training and test sets**

We use train_test_split to split the data into training and test sets.

**Create and train the linear regression model**

We create an instance of the LinearRegression model and train it with the training data.

**Make predictions and evaluate the model**

We make predictions on the test data and calculate the Mean Square Error (MSE) and the Coefficient of Determination ($R^2$).

**Viewing the data and model**

- We used Matplotlib and Seaborn to create a scatter plot showing data on prices, sizes, and number of rooms.
- We add a red line representing the linear regression model fitted to the test data.
- We customize the title and axis labels.

**Display**

Running this code will print the model evaluation metrics (MSE and $R^2$), and display a window displaying the data and the linear regression model.

The visualization allows us to see how the data on prices, sizes and number of rooms is distributed, and how the linear regression model fits this data. This can help readers better understand the relationship between variables and interpret model results.

**Conclusion**

This enhanced version of the linear regression example for predicting home prices incorporates loading data from a CSV file and visualization using Matplotlib and Seaborn. This gives readers a more realistic experience and allows them to effectively visualize data and model results. I hope this detailed explanation and the code provided are helpful in better understanding the process of linear regression modeling and data visualization.

## Project 2: Predicting the Probability of Passing an Exam

**Description:** Use **Logistic regression** to predict the probability of passing an exam based on hours of study.

**Aim:** Develop a logistic regression model using Scikit-Learn.

**Steps:**

1. Load and explore data.
2. Preprocess the data.
3. Split the data into training and test sets.
4. Train the logistic regression model.
5. Evaluate the model.
6. Test the model with new data
7. Summary

**Code:**

```
import pandas as pd
from sklearn.model_selection import train_test_split
from sklearn.linear_model import LogisticRegression
from sklearn.metrics import accuracy_score, classification_report

# Example data
data = {
    'HorasEstudio': [10, 9, 8, 7, 6, 5, 4, 3, 2, 1],
    'Approval': [1, 1, 1, 1, 1, 0, 0, 0, 0, 0]
}
df = pd.DataFrame(data)

# Independent and dependent variables
X = df[['HorasEstudio']]
y = df['I pass']

# Split into training and testing set
X_train, X_test, y_train, y_test = train_test_split(X, y, test_size=0.2, random_state=42)

# Create and train the logistic regression model
model = LogisticRegression()
model.fit(X_train, y_train)

# Predictions
y_pred = model.predict(X_test)


# Model evaluation
accuracy = accuracy_score(y_test, y_pred)
```

```
report = classification_report(y_test, y_pred)

print(f"Exactitud: {accuracy:.2f}")
print("Classification Report:")
print(report)
```

**Exit:**

Classification Report:

|  | precision | recall | f1-score | support |
|---|---|---|---|---|
| 0 | 1.00 | 1.00 | 1.00 | 1 |
| 1 | 1.00 | 1.00 | 1.00 | 1 |
| accuracy |  |  | 1.00 | 2 |
| macro avg | 1.00 | 1.00 | 1.00 | 2 |
| weighted avg | 1.00 | 1.00 | 1.00 | 2 |

Following the same steps as with the previous model, but this time with the logistic regression model.

**Prepare the new data**

Suppose we want to test the model with the following new data:

```
new_data = {
    'HorasEstudio': [6, 3, 8]
}
new_data_df = pd.DataFrame(new_data)
```

**Make predictions with new data**

```
predictions = model.predict(new_data_df)
```

**Show predictions**

```
for i, prediction in enumerate(predictions):

    print(f"Prediction for data {i+1}: {prediction}")
```

Exit:

```
Prediction for 1:1 data
Prediction for data 2:0
Prediction for 3:1 data
```

**Evaluate predictions (optional)**

If you know the actual values for the new data, you can evaluate the accuracy of the predictions:

```
actual_values = [1, 0, 1]  # Suppose these are the actual values

accuracy = accuracy_score(actual_values, predictions)
print(f"Accuracy of new data: {accuracy:.2f}")
```

Exit:

```
Accuracy on new data: 1.00
```

## 7. Summary:

In this example, the model correctly predicts whether a student will pass or fail based on study hours. The predictions are 1 (pass) for 6 and 8 hours of study, and 0 (not pass) for 3 hours of study.

Remember that this is a simple example, and in real-world problems, you will likely need more complex data and more sophisticated models to get accurate predictions.

*For use and deployment of this model follow the routine of the previous project.

# Logistic regression:
# Predicting the Probability of Passing an Exam

## Introduction

In this example, we are going to use logistic regression to predict the probability of passing an exam based on hours of study. This approach is useful in education and can help students better plan their study time.

## Data Preparation

First, we need to create a CSV file called 'exams.csv' with the study hours data and whether the student passed the exam or not. Here is the content of the file:

```
==================================================================
HorasEstudio, Approval
10,1
9,1
8,1
7,1
6,1
5,0
4,0
3,0
2,0
1,0
==================================================================
```

Then in our code, we will load this data from the CSV file using Pandas.
Code

```python
import pandas as pd
from sklearn.model_selection import train_test_split
from sklearn.linear_model import LogisticRegression
from sklearn.metrics import accuracy_score, classification_report
import matplotlib.pyplot as plt
import seaborn as sns

# Load data from a CSV file
data = pd.read_csv('examenes.csv')

# Independent and dependent variables
X = data[['HorasEstudio']]
y = data['I pass']

# Split into training and testing set
X_train, X_test, y_train, y_test = train_test_split(X, y, test_size=0.2, random_state=42)
```

```python
# Create and train the logistic regression model
model = LogisticRegression()
model.fit(X_train, y_train)

# Predictions
y_pred = model.predict(X_test)

# Model evaluation
accuracy = accuracy_score(y_test, y_pred)
report = classification_report(y_test, y_pred)

print(f"Exactitud: {accuracy:.2f}")
print("Classification Report:")
print(report)

# Visualization of the data and the model
plt.figure(figsize=(8, 6))
sns.scatterplot(x='HorasEstudio', y='Aprobo', data=data)
plt.plot(X_test['HorasEstudio'], y_pred, color='red', linewidth=2)
plt.title('Probability of Passing an Exam vs. Study Hours')
plt.xlabel('Study Hours')
plt.ylabel('Approved (0 or 1)')
plt.show()

# New data
new_data = {
    'HorasEstudio': [6, 3, 8]
}
new_data_df = pd.DataFrame(new_data)

# Make predictions with new data
predictions = model.predict(new_data_df)

# Show predictions
for i, prediction in enumerate(predictions):
    print(f"Prediction for data {i+1}: {prediction}")

# Evaluate predictions (optional)
actual_values = [1, 0, 1] # Suppose these are the actual values
accuracy = accuracy_score(actual_values, predictions)
print(f"\nAccuracy of new data: {accuracy:.2f}")
```

**Exit:**
Accuracy: 1.00
Classification Report:
```
              precision    recall  f1-score   support

           0       1.00      1.00      1.00         1
           1       1.00      1.00      1.00         1

    accuracy                           1.00         2
   macro avg       1.00      1.00      1.00         2
weighted avg       1.00      1.00      1.00         2
```

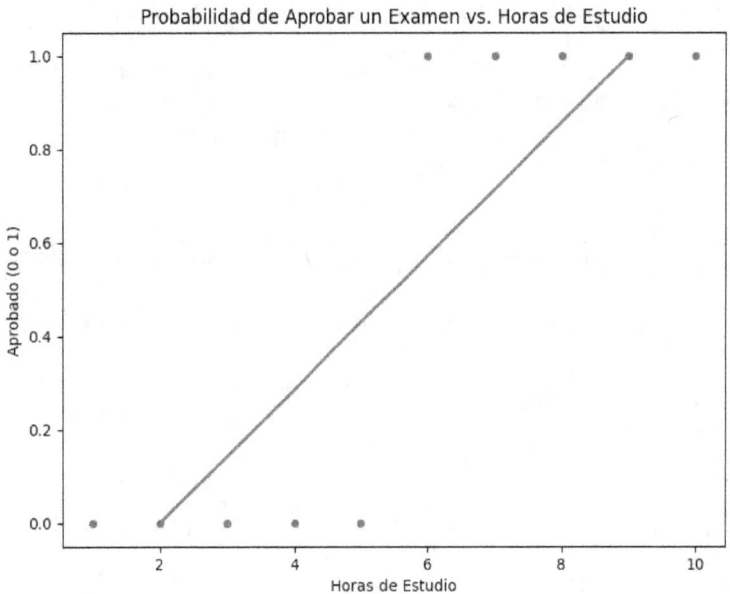

Prediction for 1:1 data
Prediction for data 2:0
Prediction for 3:1 data

Accuracy on new data: 1.00

According to predictions made with the new data:

Prediction for 1:1 data
Prediction for data 2:0
Prediction for 3:1 data
Where 1 means that the student passed the exam, and 0 means that the student did not pass.
So for the new data:
For 6 hours of study, the model predicts that the student will pass the exam (prediction = 1).
For 3 hours of study, the model predicts that the student will not pass the exam (prediction = 0).
For 8 hours of study, the model predicts that the student will pass the exam (prediction = 1).
Furthermore, if we assume that the real values are:

actual_values = [1, 0, 1]

This means that for the new data, the student actually passed with 6 and 8 hours of study, but did not pass with 3 hours of study.
By evaluating the predictions against the actual values, we obtain an accuracy of 1.00, which means that the model correctly predicted all cases.
In summary, according to the trained logistic regression model, it takes approximately 6 hours of studying or more to have a high probability of passing the exam.

## Code Explanation

**Import the necessary libraries**

We import the necessary libraries, including Matplotlib and Seaborn for visualization.

**Load data from a CSV file**

We use pd.read_csv('examenes.csv') to load the data from the CSV file 'examenes.csv'.

**Separate the independent and dependent variables**

We separate the independent variable (study hours) into X, and the dependent variable (passed or not) into y.

**Split data into training and test sets**

We use train_test_split to split the data into training and test sets.

**Create and train the logistic regression model**

We create an instance of the LogisticRegression model and train it with the training data.

**Make predictions and evaluate the model**

We make predictions on the test data and calculate the accuracy and classification report.

**Viewing the data and model**

- We used Matplotlib and Seaborn to create a scatterplot showing study hour data and whether the student passed or failed.
- We add a red line representing the logistic regression model fitted to the test data.
- We customize the title and axis labels.

**New data and predictions**

- We create a new data set with study hours.
- We use the trained model to make predictions on the new data.
- We print the predictions.

**Prediction evaluation (optional)**

If we know the actual values for the new data, we can evaluate the accuracy of the predictions using accuracy_score.

**Display**

Running this code will print the model evaluation metrics (accuracy and classification report), display a window displaying the data and the logistic regression model, and print the predictions for the new data.

The visualization allows us to see how the study hour data is distributed and whether the student passed or not, and how the logistic regression model fits this data. This can help readers better understand the relationship between variables and interpret model results.

## Conclusion

This enhanced version of the logistic regression example for predicting the probability of passing an exam incorporates loading data from a CSV file and visualization using Matplotlib and Seaborn. Additionally, it includes making predictions with new data and optionally evaluating these predictions. I hope this detailed explanation and the provided code are helpful in better understanding the logistic regression modeling process and data visualization.

## Project 3: Classification of Emails as Spam or Not Spam

**Description:** Use a **classification model** to identify emails as spam or non-spam based on the content of the email.

**Aim:** Develop a classification model using Scikit-Learn.

**Steps:**

1. Load and explore data.
2. Preprocess the data.
3. Split the data into training and test sets.
4. Train the classification model.
5. Evaluate the model
6. Test the model with new data.
7. Summary..

**Code:**

```python
import pandas as pd
from sklearn.model_selection import train_test_split
from sklearn.feature_extraction.text import TfidfVectorizer
from sklearn.naive_bayes import MultinomialNB
from sklearn.metrics import accuracy_score, classification_report

# Example data (simplified)
data = {
    'Email': ["free money", "meeting at 10", "win a prize", "project update", "lottery winner", "urgent response needed", "team lunch"],
    'Label': [1, 0, 1, 0, 1, 1, 0]
}
df = pd.DataFrame(data)

# Convert text to features
vectorizer = TfidfVectorizer()
X = vectorizer.fit_transform(df['Email'])
y = df['Label']

# Split into training and testing set
X_train, X_test, y_train, y_test = train_test_split(X, y, test_size=0.2, random_state=42)

# Create and train the Naive Bayes model
model = MultinomialNB()
model.fit(X_train, y_train)

# Predictions
y_pred = model.predict(X_test)

# Model evaluation
accuracy = accuracy_score(y_test, y_pred)
```

```
report = classification_report(y_test, y_pred)

print(f"Exactitud: {accuracy:.2f}")
print("Classification Report:")
print(report)
```

**Exit:**

```
Classification Report:
              precision    recall  f1-score   support

           0       0.00      0.00      0.00         1
           1       0.50      1.00      0.67         1

    accuracy                           0.50         2
   macro avg       0.25      0.50      0.33         2
weighted avg       0.25      0.50      0.33         2
```

6. Test the model with new data.

```
# New data
new_data = {
    'Email': ["free gift", "meeting notes", "lottery scam", "project update", "urgent reply"],
    'Label': [1, 0, 1, 0, 1]
}
new_df = pd.DataFrame(new_data)

# Vectorize the new data
new_X = vectorizer.transform(new_df['Email'])

# Make predictions
new_predictions = model.predict(new_X)

# Evaluate performance
new_accuracy = accuracy_score(new_df['Label'], new_predictions)
new_report = classification_report(new_df['Label'], new_predictions)

print(f"Accuracy on new data: {new_accuracy:.2f}")
print("Classification report on new data:")
print(new_report)
```

## 7. Summary

**Analysis of the Classification Report**

The classification report provides several metrics that allow us to evaluate the performance of the model on the new data. Let's analice each one of them:

**Precision**

- For class 0 (non-spam), the precision is 1.00, which means that when the model predicts that an email is not spam, it is correct 100% of the time.
- For class 1 (spam), the precision is 0.75, which means that when the model predicts that an email is spam, it is correct 75% of the time.

**Recall (Sensitivity)**

- For class 0 (non-spam), the recall is 0.50, which means that the model correctly identifies 50% of emails that are truly not spam.
- For class 1 (spam), the recall is 1.00, which means that the model correctly identifies 100% of the emails that are truly spam.

**F1-score**

- The F1 score is a metric that combines precision and recall, and is useful when looking for a balance between the two.
- For class 0 (non-spam), the F1-score is 0.67.
- For class 1 (spam), the F1-score is 0.86.

**Accuracy**

- The overall accuracy of the model on the new data is 0.80, which means that the model correctly classifies 80% of the emails.

**Averages**

- The macro average (macro avg) calculates the arithmetic mean of the metrics for each class, giving equal weight to each class.
- The weighted average (weighted avg) calculates the weighted average of the metrics, taking into account the number of instances of each class.

**Model evaluation**

Based on these results, we can say that the model performs well in classifying emails as spam or non-spam in the new data. However, there are some areas for improvement:

The recall for class 0 (non-spam) is relatively low (0.50), which means that the model is failing to correctly identify some emails that are not spam.

The accuracy for class 1 (spam) is 0.75, indicating that the model may be misclassifying some emails as spam when they are not.

Overall, the model appears to be well trained, but could benefit from additional tuning or exploring other classification techniques to improve its performance in problematic cases.

# Add data from a CSV file and display

We can modify the code to read the data from a CSV file and add visualizations using the Matplotlib and Seaborn libraries. Here is the updated code:

Before installing the Matplotlib and Seaborn modules

**pip install matplotlib seaborn**

Use Anaconda's JupyterLab for its greater flexibility or in the visual studio code itself. In both environments the result is the same.

**Code:**

```
import pandas as pd
from sklearn.model_selection import train_test_split
from sklearn.feature_extraction.text import TfidfVectorizer
from sklearn.naive_bayes import MultinomialNB
from sklearn.metrics import accuracy_score, classification_report
import matplotlib.pyplot as plt
import seaborn as sns

# Load data from a CSV file
data = pd.read_csv('./emails.csv')

# Convert text to features
vectorizer = TfidfVectorizer()
X = vectorizer.fit_transform(data['Email'])
y = data['Label']

# Split into training and testing set
X_train, X_test, y_train, y_test = train_test_split(X, y, test_size=0.2, random_state=42)

# Create and train the Naive Bayes model
model = MultinomialNB()
model.fit(X_train, y_train)

# Predictions
y_pred = model.predict(X_test)

# Model evaluation
accuracy = accuracy_score(y_test, y_pred)
report = classification_report(y_test, y_pred)

print(f"Exactitud: {accuracy:.2f}")
print("Classification Report:")
print(report)

# Visualization of the confusion matrix
cm = pd.crosstab(y_test, y_pred, rownames=['Actual'], colnames=['Prediction'])
plt.figure(figsize=(6, 4))
sns.heatmap(cm, annot=True, cmap='Blues', fmt='g')
plt.title('Confusion Matrix')
plt.xlabel('Prediction')
```

```
plt.ylabel('Actual')
plt.show()
```

Exit:

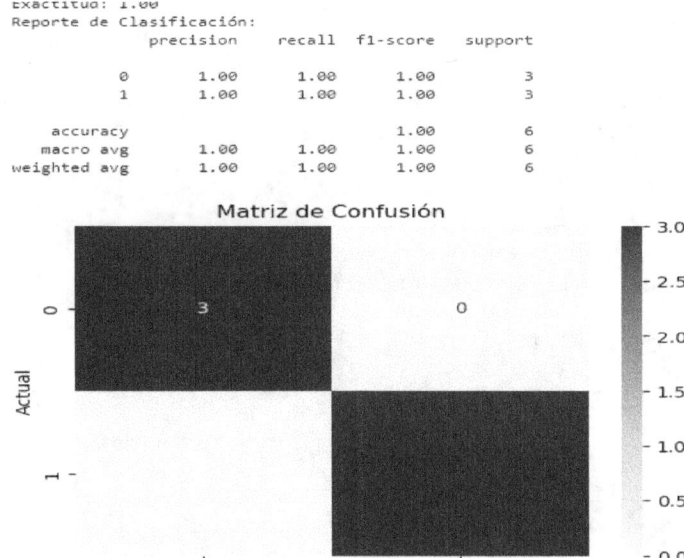

Explanation of the changes:
Load data from a CSV file
We use pd.read_csv('emails.csv') to load the data from a CSV file called 'emails.csv'.
We assume that the CSV file has columns named 'Email' and 'Label'.
Import visualization libraries
We import matplotlib.pyplot and seaborn to create visualizations.
Visualization of the confusion matrix
After evaluating the model, we created a confusion matrix using pd.crosstab.
We use Seaborn's sns.heatmap to visualize the confusion matrix as a heatmap.
We customized the figure size, title, axis labels, and annotation format.
Make sure you have a CSV file called 'emails.csv' in the same directory as your Python script, with the 'Email' and 'Label' columns. The CSV file must contain the emails and their corresponding tags (spam or non-spam).
Running this code will print the accuracy and classification report, and will also display a window displaying the confusion matrix.

Here you have a CSV file called 'emails.csv' with fake email data and their corresponding labels (spam or non-spam). I have included more records to have a larger data set.

**emails.cvs**
==================================================================
```
Email,Label
"Get rich quick! Click here!",1
"Meeting agenda for tomorrow",0
"Earn money from home today!",1
"Project update: Phase 1 completed",0
"Urgent: Your account has been compromised",1
"Team lunch at 12:30 PM",0
"You've won a free vacation!",1
"Monthly sales report",0
"Congratulations! You're a winner!",1
"Important: Software update required",0
"Earn extra cash with this simple trick",1
"Weekly status meeting reminder",0
"Claim your prize now!",1
"New product launch details",0
"Urgent action required: Your account will be suspended",1
"Company-wide announcement",0
"Make millions with this secret formula!",1
"Meeting minutes from last week",0
"Attention: Your order has been shipped",0
"Earn passive income today!",1
"Important security update",0
"You've been selected for a special offer!",1
"Quarterly financial report",0
"Urgent: Your password has been reset",1
"Team-building event details",0
"Limited time offer: Buy now and save!",1
"Project kickoff meeting agenda",0
"Congratulations! You've won a free gift card!",1
"Company policy update",0
"Earn extra cash by taking surveys!",1
```
==================================================================

This CSV file contains 30 records, with fake emails and random spam (1) or non-spam (0) labels. You can use this file directly in your code to load the data and train the classification model. Please remember that these data are simply examples and do not represent actual emails. In a real situation, you should use a larger and more representative data set to train your model effectively.

# Summary of the Projects in this Section

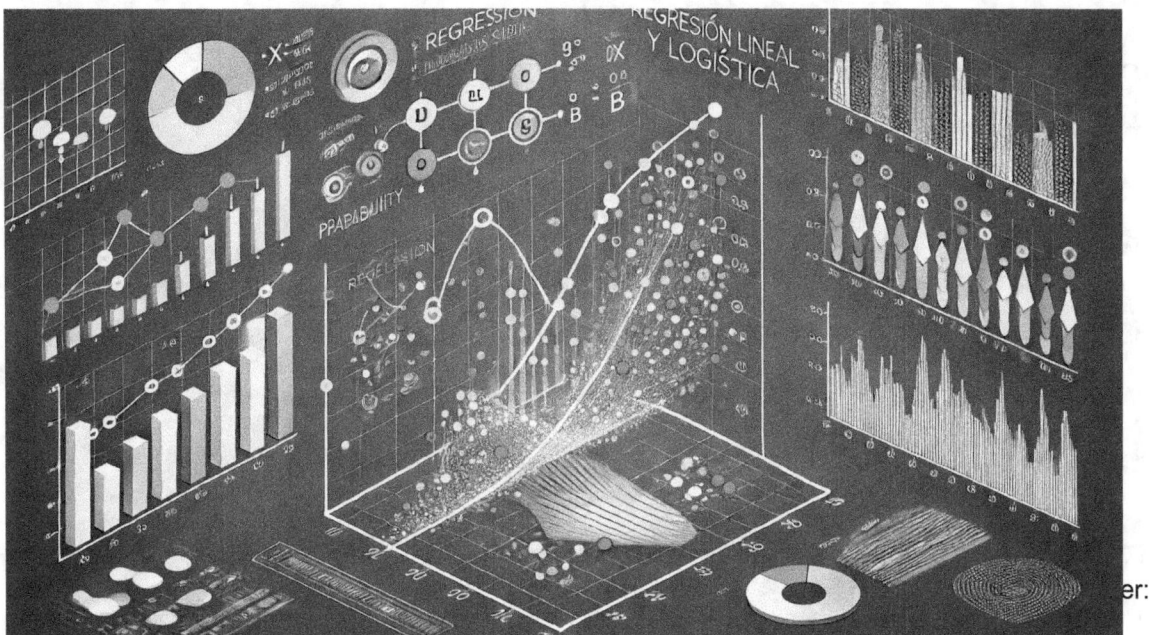

## Project 1: Home Price Prediction

In this project, we use linear regression to predict home prices based on their size and number of bedrooms. We follow the following steps:

- We load data on size, number of rooms and home prices from a CSV file.
- We split the data into training and test sets.
- We train the linear regression model using the training set.
- We evaluate the performance of the model on the test set using metrics such as Mean Square Error (MSE) and Coefficient of Determination ($R^2$).
- We visualize the data and the fitted linear regression model.
- We make price predictions with new data on size and number of rooms.

This project allowed us to understand how to build and evaluate a linear regression model to predict a continuous variable (price) as a function of independent variables (size and number of rooms).

## Project 2: Predicting the Probability of Passing an Exam

In this project, we use logistic regression to predict the probability of a student passing an exam based on hours of study. We follow the following steps:

- We load the data on study hours and whether or not the student passed the exam from a CSV file.
- We split the data into training and test sets.
- We train the logistic regression model using the training set.
- We evaluate the performance of the model on the test set using metrics such as accuracy and classification reporting.
- We visualize the data and the fitted logistic regression model.
- We make predictions with new study hour data and evaluate the results.

This project allowed us to understand how to build and evaluate a logistic regression model to predict a binary variable (pass or fail) based on an independent variable (study hours).

## Project 3: Classification of Emails as Spam or Not Spam

In this project, we use a classification model based on the Naive Bayes algorithm to identify emails as spam or non-spam based on their content. We follow the following steps:

- We load and prepare email data and their corresponding labels.
- We split the data into training and test sets.
- We train the Naive Bayes classification model using the training set.
- We evaluate the performance of the model on the test set using metrics such as precision, recall, and F1 score.
- We make predictions with new email data and evaluate the results.
- We visualize the confusion matrix to better understand the performance of the model.

This project allowed us to understand how to build and evaluate a classification model for a real-world problem, such as spam filtering.

---

These three projects gave us hands-on experience developing machine learning models using Scikit-Learn, as well as visualizing data and interpreting results. We hope this section has been helpful in understanding the concepts and techniques involved in data modeling and prediction.

# Decision Trees and Random Forests

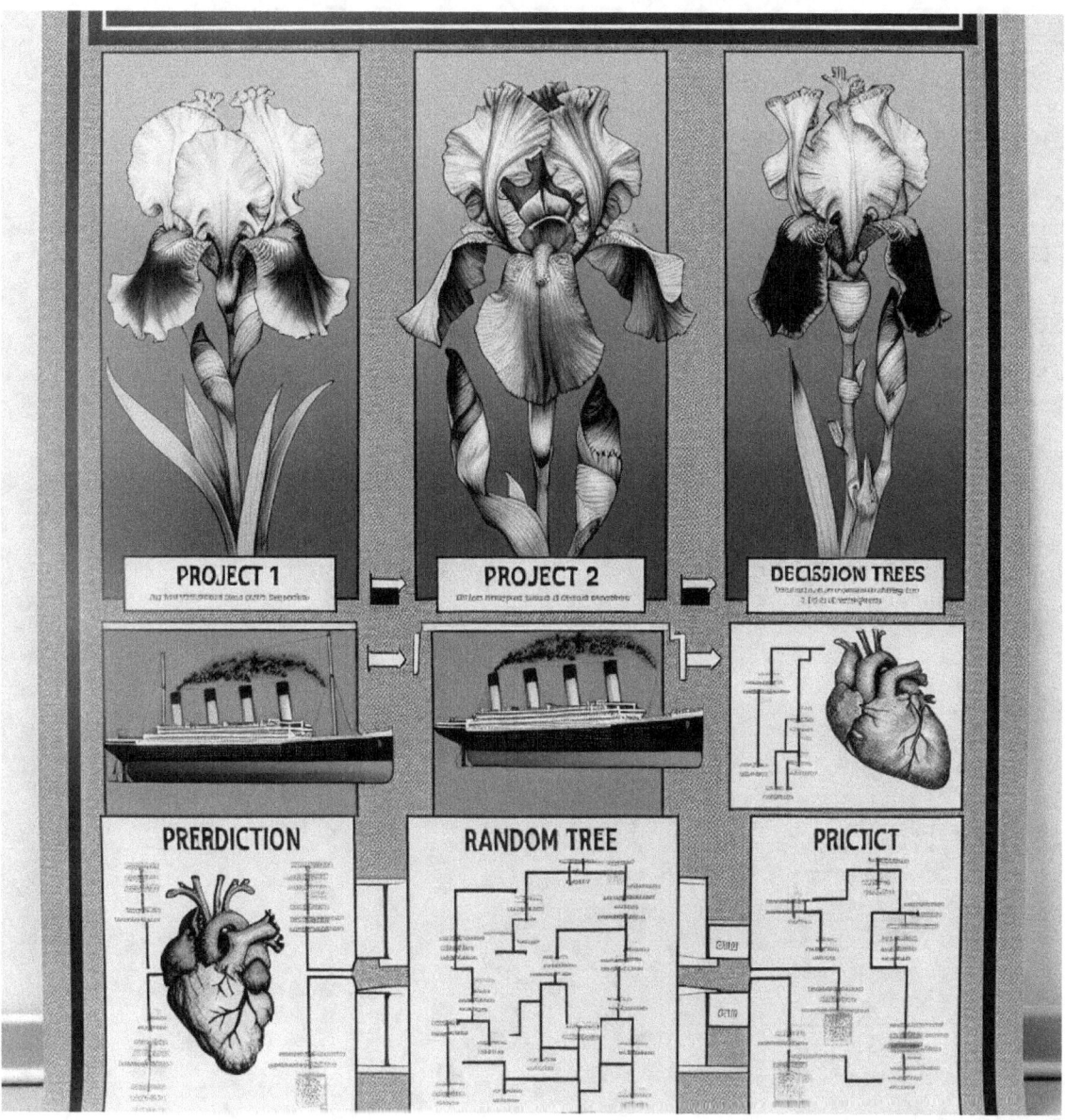

## Project 1: Classification of Iris Flowers

Description: Use decision trees to classify Iris flower species based on characteristics such as length and width of the sepal and petal.

Objective: Develop a decision tree model using Scikit-Learn.

Steps:

1. Load and explore data.
2. Preprocess the data.
3. Split the data into training and test sets.
4. Train the decision tree model.
5. Evaluate the model.

Code:

```python
import pandas as pd
from sklearn.model_selection import train_test_split
from sklearn.tree import DecisionTreeClassifier
from sklearn.metrics import accuracy_score, classification_report
from sklearn.datasets import load_iris

# Load the Iris dataset
iris = load_iris()
X = iris.data
y = iris.target

# Split into training and testing set
X_train, X_test, y_train, y_test = train_test_split(X, y, test_size=0.2, random_state=42)

# Create and train the decision tree model
model = DecisionTreeClassifier()
model.fit(X_train, y_train)

# Predictions
y_pred = model.predict(X_test)

# Model evaluation
accuracy = accuracy_score(y_test, y_pred)
report = classification_report(y_test, y_pred, target_names=iris.target_names)

print(f"Exactitud: {accuracy:.2f}")
print("Classification Report:")
print(report)
```

**Exit:**

Accuracy: 1.00

Classification Report:

|              | precision | recall | f1-score | support |
|--------------|-----------|--------|----------|---------|
| silky        | 1.00      | 1.00   | 1.00     | 10      |
| versicolor   | 1.00      | 1.00   | 1.00     | 9       |
| virginica    | 1.00      | 1.00   | 1.00     | 11      |
| accuracy     |           |        | 1.00     | 30      |
| macro avg    | 1.00      | 1.00   | 1.00     | 30      |
| weighted avg | 1.00      | 1.00   | 1.00     | 30      |

## Project 1: Classification of Iris Flowers with Visualization and Test

### Introduction

In this project, we will use the famous Iris flower dataset to build a decision tree-based classification model. The goal is to predict the species of an Iris flower (Setosa, Versicolor or Virginica) based on characteristics such as the length and width of the sepals and petals.

### Data Preparation

Fortunately, the Iris dataset is built into Scikit-learn, so we don't need to load the data from an external file. We will use the load_iris() function to load the data directly.

Code

```python
import pandas as pd
from sklearn.model_selection import train_test_split
from sklearn.tree import DecisionTreeClassifier
from sklearn.metrics import accuracy_score, classification_report
from sklearn.datasets import load_iris

# Load the Iris dataset
iris = load_iris()
X = iris.data
y = iris.target

# Split into training and testing set
X_train, X_test, y_train, y_test = train_test_split(X, y, test_size=0.2, random_state=42)

# Create and train the decision tree model
model = DecisionTreeClassifier()
model.fit(X_train, y_train)

# Predictions
y_pred = model.predict(X_test)

# Model evaluation
accuracy = accuracy_score(y_test, y_pred)
report = classification_report(y_test, y_pred, target_names=iris.target_names)

print(f"Exactitud: {accuracy:.2f}")
print("Classification Report:")
print(report)
```

**Exit:**
```
Accuracy: 1.00
Classification Report:
              precision    recall  f1-score   support

       silky       1.00      1.00      1.00        10
  versicolor       1.00      1.00      1.00         9
   virginica       1.00      1.00      1.00        11

    accuracy                           1.00        30
   macro avg       1.00      1.00      1.00        30
weighted avg       1.00      1.00      1.00        30
```

## Code Explanation

### Import the necessary libraries

We import the necessary libraries from Scikit-learn, including load_iris to load the dataset.

### Load the Iris dataset

- We use load_iris() to load the Iris dataset.
- The data is split into X (features) and y (class labels).

### Split data into training and test sets

- We use train_test_split to split the data into training and test sets.
- In this case, 20% of the data will be used for testing (test_size=0.2).

### Create and train the decision tree model

- We create an instance of the DecisionTreeClassifier model.
- We train the model with the training data using the fit method.

### Make predictions and evaluate the model

- We make predictions on the test data with the predict method.
- We calculate the accuracy using accuracy_score.
- We generate a detailed classification report with classification_report, including the class names (Setosa, Versicolor, Virginica).

### Print the results

We print the accuracy and classification report to evaluate the performance of the model.

## Visualization (Optional)

If you want to visualize the trained decision tree, you can use the graphviz library and the export_graphviz method of Scikit-learn. Here is an example of how to do it:

Install the graphviz module

**`pip install graphviz`**

*Use:*
*Make this project in visual studio code.*

```
import graphviz

# Display the decision tree
dot_data = tree.export_graphviz(model, out_file=None,

feature_names=iris.feature_names,
                                class_names=iris.target_names,
                                filled=True, rounded=True)

graph = graphviz.Source(dot_data)
graph.render("iris_tree")
```

This code will generate a file **iris_tree.pdf** which contains a graphical representation of the trained decision tree.

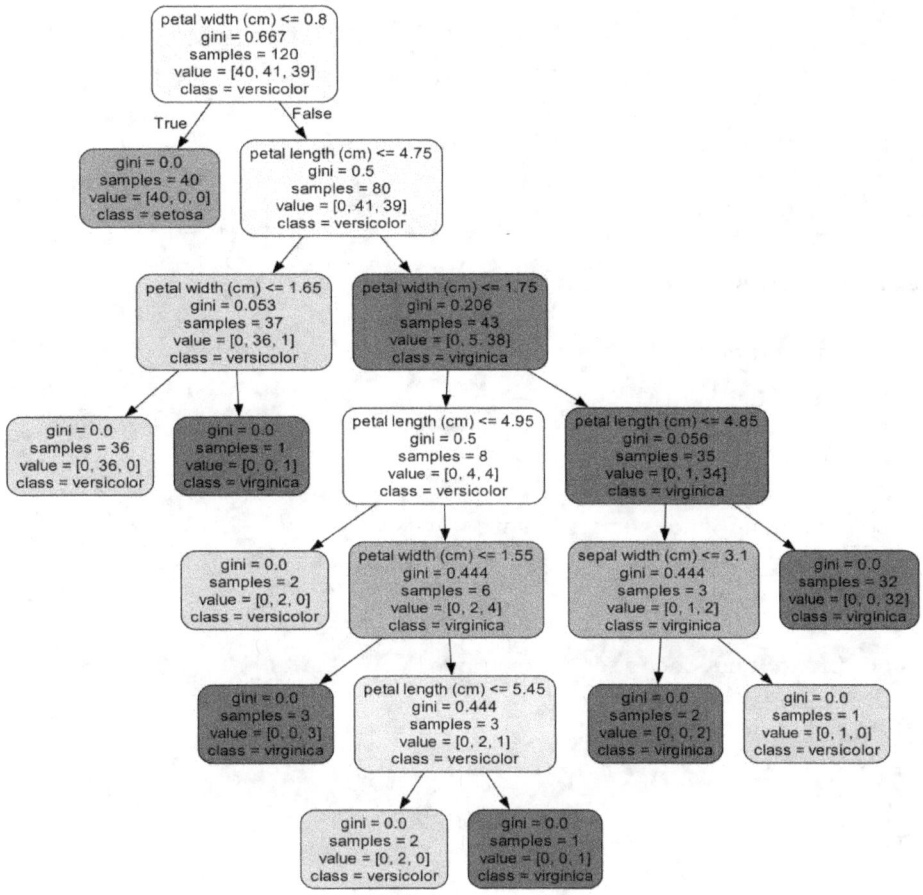

Let's go back to the code from the Iris Flower Classification project and add some lines to print the predictions, the actual values, and the confusion matrix. Here is the updated code:

```python
import pandas as pd
from sklearn.model_selection import train_test_split
from sklearn.tree import DecisionTreeClassifier
from sklearn.metrics import accuracy_score, classification_report
from sklearn.datasets import load_iris
from sklearn import tree
import graphviz
from sklearn.metrics import confusion_matrix

# Load the Iris dataset
iris = load_iris()
X = iris.data
y = iris.target

# Split into training and testing set
X_train, X_test, y_train, y_test = train_test_split(X, y, test_size=0.2, random_state=42)

# Create and train the decision tree model
model = DecisionTreeClassifier()
model.fit(X_train, y_train)

# Predictions
y_pred = model.predict(X_test)

# Model evaluation
accuracy = accuracy_score(y_test, y_pred)
report = classification_report(y_test, y_pred, target_names=iris.target_names)

print(f"Exactitud: {accuracy:.2f}")
print("Classification Report:")
print(report)

# Print predictions and actual values
print("\nPredictions: ", y_pred)
print("Actual values: ", y_test)

# Confusion matrix
conf_matrix = confusion_matrix(y_test, y_pred)
print("\nConfusion matrix:")
print(conf_matrix)
```

**In this code, we have added the following lines:**

```
from sklearn import tree
import graphviz
from sklearn.metrics import confusion_matrix

...

# Print predictions and actual values
print("\nPredictions: ", y_pred)
print("Actual values: ", y_test)

# Confusion matrix
conf_matrix = confusion_matrix(y_test, y_pred)
print("\nConfusion matrix:")
print(conf_matrix)
```

Now, when you run this code, you will get the accuracy and classification report, followed by the predictions and actual values for the test set, and finally the confusion matrix.

You will be able to analyze the predictions and compare them with the real values. Additionally, the confusion matrix will give you valuable information about the model's true positives, false positives, true negatives, and false negatives.

This way, you will have a more complete analysis of the results of the Iris Flower Classification model, which will allow you to better understand its performance.

One of the main goals of machine learning is to develop models that can make accurate and reliable predictions. However, it is not enough to just train a model and get an accuracy score. It is crucial to analyze and understand the results in more detail.

By printing out the individual predictions, comparing them to the actual values, and examining the confusion matrix, we get a clearer view of how the model is performing in different cases. This allows us to identify patterns, strengths and weaknesses of the model, which is essential to be able to improve or adjust it as necessary.

Many introductory books and resources focus on the basic concepts and implementation of algorithms, but often do not delve into these results analysis techniques. However, in real practice, these tools are essential for evaluating and optimizing the performance of machine learning models.

These more detailed and practical approaches will give you a stronger understanding and better prepare you to face real challenges in the field of machine learning.

We will continue to share more advanced techniques and approaches as we progress with our projects. My goal is to ensure that you gain in-depth, practical knowledge that allows you to apply these skills effectively in real-world situations.

**Exit:**

```
Accuracy: 1.00
Classification Report:
              precision    recall  f1-score   support

       silky       1.00      1.00      1.00        10
  versicolor       1.00      1.00      1.00         9
   virginica       1.00      1.00      1.00        11

    accuracy                           1.00        30
   macro avg       1.00      1.00      1.00        30
weighted avg       1.00      1.00      1.00        30


Predictions:  [1 0 2 1 1 0 1 2 1 1 2 0 0 0 0 1 2 1 1 2 0 2 0 2 2 2 2 2 0 0]
Actual values: [1 0 2 1 1 0 1 2 1 1 2 0 0 0 0 1 2 1 1 2 0 2 0 2 2 2 2 2 0 0]

Confusion matrix:
[[10  0  0]
 [ 0  9  0]
 [ 0  0 11]]
```

## Let's analyze each part:

### Accuracy and Classification Report:

> The accuracy of the model is 1.00, which means that 100% of the predictions were correct on the test set. The classification report also shows a precision, recall and F1 score of 1.00 for each of the three classes (setosa, versicolor and virginica), which is an exceptional result.

### Predictions and Real Values:

> By comparing the predictions with the actual values, we can see that the model correctly classified all instances in the test set. Each prediction exactly matches the corresponding actual value.

**Confusion Matrix:**

The confusion matrix confirms the excellent results of the model. The main diagonal shows the true positives for each class (10 for setosa, 9 for versicolor, and 11 for virginica), while the rest of the values are zeros, indicating that there were no false positives or false negatives.

In summary, these results indicate that the decision tree model trained on the Iris dataset achieved perfect performance in classifying all three flower species in the test set.

This is an exceptional and rare result in machine learning problems, where a certain level of error or confusion in predictions is generally expected. However, in this particular case, the model appears to have perfectly captured the distinctive patterns and characteristics of each flower species.

It is worth mentioning that the Iris data set is relatively small and well-defined, which can facilitate the classification task for an algorithm such as the decision tree. However, this result is an excellent example of how a well-trained and tested model can achieve optimal performance on a specific problem.

## How to Test this Model?

You can test the decision tree model with data from any flower to classify which species it belongs to.

The model trained in this project uses the following features to perform classification:

Sepal length

Sepal width

Petal length

Petal Width

Therefore, if you give me the numerical values of those four characteristics for a particular flower, I can pass them to the model and get the prediction of the species it belongs to (setosa, versicolor or virginica).

To make the prediction, you can follow these steps:

Create a Pandas DataFrame with a row containing the values of the four characteristics for the flower you want to classify.

```
import pandas as pd

# Example with dummy values
nueva_flor = pd.DataFrame({'Sepal Length': [5.8],
                           'Sepal Width': [2.7],
                           'Petal Length': [5.1],
                           'Petal Width': [1.9]})
```

Use the predict method of the trained model and pass it the DataFrame new_flower as an argument:

```
prediction = model.predict(new_flower)
```

The prediction will be an integer representing the class (0 for setosa, 1 for versicolor, 2 for virginica). You can map that number to the corresponding class label:

```
species = ['setosa', 'versicolor', 'virginica']
predicted_species = species[prediction[0]]
print(f"The flower belongs to the species: {predicted_species}")
```

Simply replace the example values in new_flower with the actual measurements of the flower you want to classify, and the model will make the corresponding prediction.

Remember that the accuracy of the prediction will depend on how well the model has been trained and the quality of the data used during training. But this is the general way to use the model to classify new flowers based on their characteristics.

**Exit:**

```
The flower belongs to the species: virginica
```

## Conclusion

In this project, we have built a decision tree based classification model using the Iris dataset. We have followed the standard steps of loading data, splitting into training and testing sets, training the model, making predictions, and evaluating performance.

Decision trees are a popular and easy-to-interpret machine learning algorithm, making them well-suited for classification problems like this one. However, they also have limitations, such as overfitting in some cases.

You can explore this project further by tuning the hyperparameters of the decision tree, trying different data preprocessing techniques, or even combining multiple models in an ensemble approach to improve performance.

## Project 2: Titanic Survival Prediction

**Description:** Use decision trees to predict the survival of Titanic passengers based on characteristics such as age, sex, and class.

**Aim:** Develop a decision tree model using Scikit-Learn.

**Steps:**

1. Load and explore data.
2. Preprocess the data.
3. Split the data into training and test sets.
4. Train the decision tree model.
5. Evaluate the model.

Code:

```
import pandas as pd
from sklearn.model_selection import train_test_split
from sklearn.tree import DecisionTreeClassifier
from sklearn.metrics import accuracy_score, classification_report

# Load the Titanic dataset
url = "https://web.stanford.edu/class/archive/cs/cs109/cs109.1166/stuff/titanic.csv"
df = pd.read_csv(url)

# Preprocess the data
df['Sex'] = df['Sex'].map({'male': 0, 'female': 1})
df['Age'].fillna(df['Age'].mean(), inplace=True)
X = df[['Pclass', 'Sex', 'Age']]
y = df['Survived']

# Split into training and testing set
X_train, X_test, y_train, y_test = train_test_split(X, y, test_size=0.2, random_state=42)

# Create and train the decision tree model
model = DecisionTreeClassifier()
model.fit(X_train, y_train)

# Predictions
y_pred = model.predict(X_test)

# Model evaluation
accuracy = accuracy_score(y_test, y_pred)
report = classification_report(y_test, y_pred)
```

```
print(f"Exactitud: {accuracy:.2f}")
print("Classification Report:")
print(report)
```

**Exit:**
Accuracy: 0.77
Classification Report:

|              | precision | recall | f1-score | support |
|--------------|-----------|--------|----------|---------|
| 0            | 0.77      | 0.89   | 0.83     | 111     |
| 1            | 0.76      | 0.57   | 0.65     | 67      |
|              |           |        |          |         |
| accuracy     |           |        | 0.77     | 178     |
| macro avg    | 0.77      | 0.73   | 0.74     | 178     |
| weighted avg | 0.77      | 0.77   | 0.76     | 178     |

## Project 2: Titanic Survival Prediction

### Introduction

In this project, we will use the famous Titanic data set to build a classification model based on decision trees. The goal is to predict whether or not a passenger survived the sinking of the Titanic based on characteristics such as age, gender, and the class in which they were traveling.

### Data Preparation

In this case, we will load the data directly from a URL using Pandas. The data set contains information about the passengers of the Titanic, including their age, gender, class, and whether or not they survived the sinking.

We will perform some data preprocessing steps, such as encoding the Sex variable as numeric and filling in missing Age values with the mean.

### Code

```python
import pandas as pd
from sklearn.model_selection import train_test_split
from sklearn.tree import DecisionTreeClassifier
from sklearn.metrics import accuracy_score, classification_report

# Load the Titanic dataset
url = "https://web.stanford.edu/class/archive/cs/cs109/cs109.1166/stuff/titanic.csv"
df = pd.read_csv(url)

# Preprocess the data
df['Sex'] = df['Sex'].map({'male': 0, 'female': 1})
df['Age'].fillna(df['Age'].mean(), inplace=True)
X = df[['Pclass', 'Sex', 'Age']]
y = df['Survived']

# Split into training and testing set
X_train, X_test, y_train, y_test = train_test_split(X, y, test_size=0.2, random_state=42)

# Create and train the decision tree model
model = DecisionTreeClassifier()
model.fit(X_train, y_train)

# Predictions
y_pred = model.predict(X_test)

# Model evaluation
accuracy = accuracy_score(y_test, y_pred)
report = classification_report(y_test, y_pred)
```

```
print(f"Exactitud: {accuracy:.2f}")
print("Classification Report:")
print(report)
```

**Exit:**

```
Accuracy: 0.77
Classification Report:
              precision    recall  f1-score   support

           0       0.77      0.89      0.83       111
           1       0.76      0.57      0.65        67

    accuracy                           0.77       178
   macro avg       0.77      0.73      0.74       178
weighted avg       0.77      0.77      0.76       178
```

## Code Explanation

Import the necessary libraries

We import the necessary libraries from Pandas and Scikit-learn.

### Load Titanic Dataset

We use pd.read_csv to load the data directly from the provided URL.

### Preprocess the data

- We coded the Sex variable as numeric (0 for 'male', 1 for 'female').
- We fill the missing Age values with the mean using fillna.
- We select the independent variables (Pclass, Sex, Age) in X and the dependent variable (Survived) in y.

### Split data into training and test sets

- We use train_test_split to split the data into training and test sets.
- In this case, 20% of the data will be used for testing (test_size=0.2).

### Create and train the decision tree model

- We create an instance of the DecisionTreeClassifier model.
- We train the model with the training data using the fit method.

### Make predictions and evaluate the model

- We make predictions on the test data with the predict method.
- We calculate the accuracy using accuracy_score.
- We generate a detailed classification report with classification_report.

**Print the results**

We print the accuracy and classification report to evaluate the performance of the model.

**Visualization (Optional)**

As in the previous project, you can use the graphviz library and Scikit-learn's export_graphviz method to visualize the trained decision tree. Here is an example of how to do it:

```
import graphviz

# Display the decision tree
dot_data = tree.export_graphviz(model, out_file=None,
                                feature_names=['Pclass', 'Sex', 'Age'],
                                class_names=['No Survived', 'Survived'],
                                filled=True, rounded=True)

graph = graphviz.Source(dot_data)
graph.render("titanic_tree")
```

This code will generate a file **titanic_tree.pdf** which contains a graphical representation of the trained decision tree.

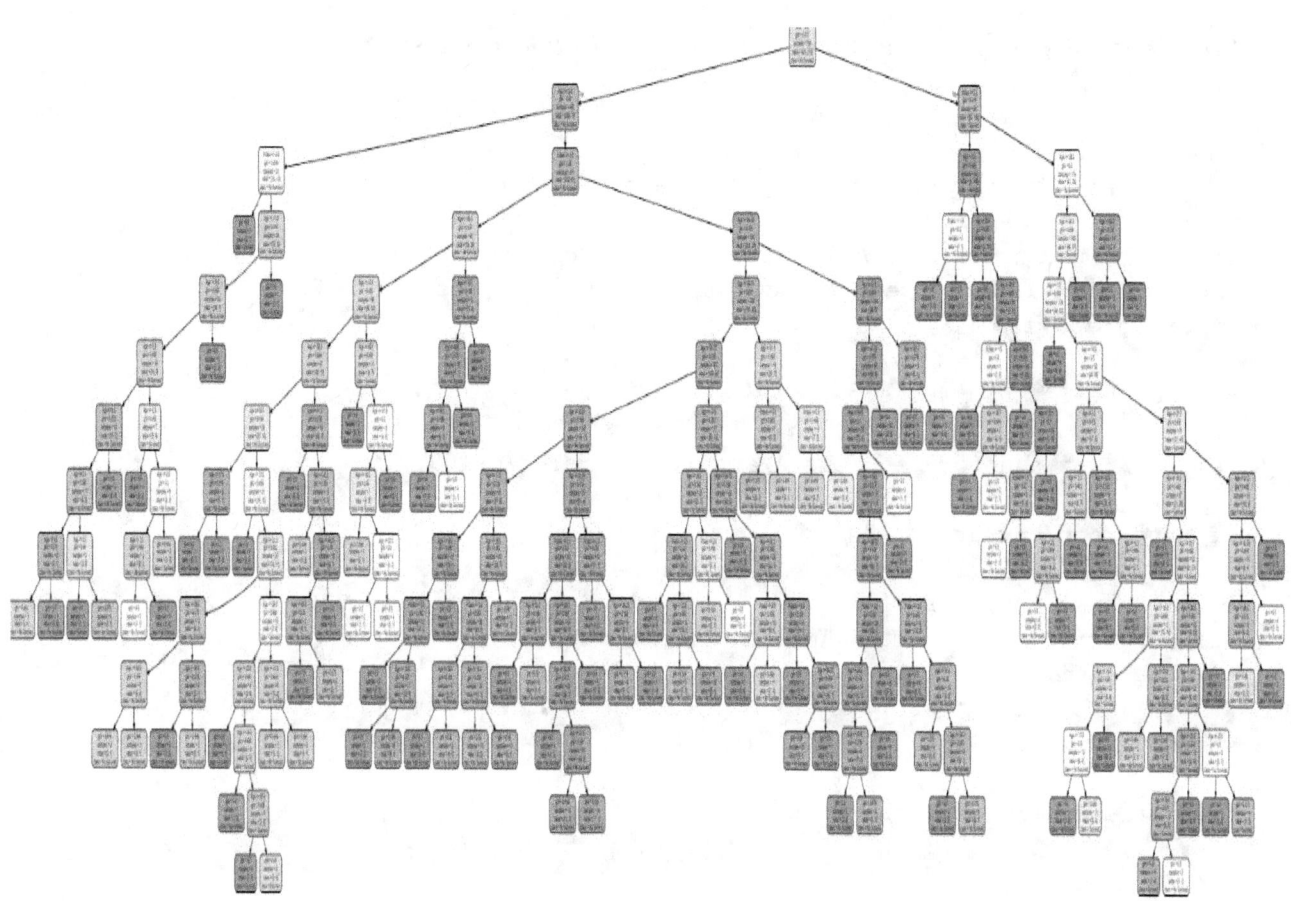

**Complete Code:**

*Note: Remember for this type of project to use visual studio code It is.*

```python
import pandas as pd
from sklearn.model_selection import train_test_split
from sklearn.tree import DecisionTreeClassifier
from sklearn.metrics import accuracy_score, classification_report
from sklearn import tree   # Added line to import tree module
import graphviz

# Load the Titanic dataset
url = "https://web.stanford.edu/class/archive/cs/cs109/cs109.1166/stuff/titanic.csv"
df = pd.read_csv(url)

# Preprocess the data
df['Sex'] = df['Sex'].map({'male': 0, 'female': 1})
df['Age'].fillna(df['Age'].mean(), inplace=True)
X = df[['Pclass', 'Sex', 'Age']]
y = df['Survived']

# Split into training and testing set
X_train, X_test, y_train, y_test = train_test_split(X, y, test_size=0.2, random_state=42)

# Create and train the decision tree model
model = DecisionTreeClassifier()
model.fit(X_train, y_train)

# Predictions
y_pred = model.predict(X_test)

# Model evaluation
accuracy = accuracy_score(y_test, y_pred)
report = classification_report(y_test, y_pred)

print(f"Exactitud: {accuracy:.2f}")
print("Classification Report:")
print(report)
```

```python
# Display the decision tree
dot_data = tree.export_graphviz(model, out_file=None,
                                feature_names=['Pclass', 'Sex', 'Age'],
                                class_names=['No Survived', 'Survived'],
                                filled=True, rounded=True)

graph = graphviz.Source(dot_data)
graph.render("titanic_tree")
```

## Conclusion

In this project, we have built a decision tree-based classification model using the Titanic dataset. We have followed the standard steps of loading data, preprocessing, splitting into training and test sets, training the model, making predictions, and evaluating performance.

Decision trees are an interpretable and easy-to-understand algorithm, making them suitable for classification problems like this. However, they also have limitations, such as overfitting in some cases.

You can explore this project further by tuning the hyperparameters of the decision tree, trying different data preprocessing techniques, or even combining multiple models in an ensemble approach to improve performance.

Additionally, you can try to visualize the trained decision tree to get a better understanding of the decision rules and features most important for predicting survival on the Titanic.

## Project 3: Heart Disease Prediction

**Description:** Using random forests to predict heart disease based on medical characteristics.

**Aim:** Develop a random forest model using Scikit-Learn.

**Steps:**

1. Load and explore data.
2. Preprocess the data.
3. Split the data into training and test sets.
4. Train the random forest model.
5. Evaluate the model.

**Code:**

```
import pandas as pd
from sklearn.model_selection import train_test_split
from sklearn.ensemble import RandomForestClassifier
from sklearn.metrics import accuracy_score, classification_report

# Load heart disease dataset
url = "https://archive.ics.uci.edu/ml/machine-learning-databases/heart-disease/processed.cleveland.data"
columns = ['age', 'sex', 'cp', 'trestbps', 'chol', 'fbs', 'restecg', 'thalach', 'exang', 'oldpeak', 'slope', 'ca', 'thal', 'target']
df = pd.read_csv(url, names=columns)

# Preprocess the data
df.replace('?', -1, inplace=True)
df = df.apply(pd.to_numeric)
X = df.drop('target', axis=1)
y = df['target']

# Split into training and testing set
X_train, X_test, y_train, y_test = train_test_split(X, y, test_size=0.2, random_state=42)

# Create and train the random forest model
model = RandomForestClassifier(n_estimators=100, random_state=42)
model.fit(X_train, y_train)

# Predictions
y_pred = model.predict(X_test)

# Model evaluation
accuracy = accuracy_score(y_test, y_pred)
report = classification_report(y_test, y_pred)
```

```
print(f"Exactitud: {accuracy:.2f}")
print("Classification Report:")
print(report)
```

**Exit:**

Accuracy: 0.49

Classification Report:

|   | precision | recall | f1-score | support |
|---|---|---|---|---|
| 0 | 0.72 | 0.97 | 0.82 | 29 |
| 1 | 0.10 | 0.08 | 0.09 | 12 |
| 2 | 0.14 | 0.11 | 0.12 | 9 |
| 3 | 0.00 | 0.00 | 0.00 | 7 |
| 4 | 0.00 | 0.00 | 0.00 | 4 |
| accuracy |  |  | 0.49 | 61 |
| macro avg | 0.19 | 0.23 | 0.21 | 61 |
| weighted avg | 0.38 | 0.49 | 0.43 | 61 |

## Project 3: Heart Disease Prediction following the same detailed methodology that we have used in previous projects.

```python
import pandas as pd
from sklearn.model_selection import train_test_split
from sklearn.ensemble import RandomForestClassifier
from sklearn.metrics import accuracy_score, classification_report

# Load heart disease dataset
url = "https://archive.ics.uci.edu/ml/machine-learning-databases/heart-disease/processed.cleveland.data"
columns = ['age', 'sex', 'cp', 'trestbps', 'chol', 'fbs', 'restecg',
'thalach', 'exang', 'oldpeak', 'slope', 'ca', 'thal', 'target']
df = pd.read_csv(url, names=columns)

# Preprocess the data
df.replace('?', -1, inplace=True)
df = df.apply(pd.to_numeric)
X = df.drop('target', axis=1)
y = df['target']

# Split into training and testing set
X_train, X_test, y_train, y_test = train_test_split(X, y, test_size=0.2, random_state=42)

# Create and train the random forest model
model = RandomForestClassifier(n_estimators=100, random_state=42)
model.fit(X_train, y_train)

# Predictions
y_pred = model.predict(X_test)

# Model evaluation
accuracy = accuracy_score(y_test, y_pred)
report = classification_report(y_test, y_pred)

print(f"Exactitud: {accuracy:.2f}")
print("Classification Report:")
print(report)

# Print predictions and actual values
print("\nPredictions: ", y_pred)
print("Actual values: ", y_test)

# Confusion matrix
from sklearn.metrics import confusion_matrix
conf_matrix = confusion_matrix(y_test, y_pred)
```

```python
print("\nConfusion matrix:")
print(conf_matrix)
```

**Exit:**

```
Accuracy: 0.49
Classification Report:
              precision    recall  f1-score   support

           0       0.72      0.97      0.82        29
           1       0.10      0.08      0.09        12
           2       0.14      0.11      0.12         9
           3       0.00      0.00      0.00         7
           4       0.00      0.00      0.00         4

    accuracy                           0.49        61
   macro avg       0.19      0.23      0.21        61
weighted avg       0.38      0.49      0.43        61


Predictions: [0 1 2 0 1 3 2 1 0 0 0 0 1 1 1 0 0 0 2 0 0 0 3 0 3 0 0 2 0 1
 0 0 0 0 0 0 2
 0 1 0 2 0 1 0 0 3 0 0 0 0 0 0 0 0 0 3 0 0 2 1 0 0]
Actual values: 179    0
228    3
111    1
246    2
60     2
      ..
249    0
104    3
300    3
193    2
184    1
Name: target, Length: 61, dtype: int64

Confusion matrix:
[[28  0  1  0  0]
 [ 6  1  3  2  0]
 [ 5  1  1  2  0]
 [ 0  6  1  0  0]
 [ 0  2  1  1  0]]
```

## Code Explanation

**Import the necessary libraries**

- We imported the necessary libraries from Pandas and Scikit-learn, including RandomForestClassifier for the random forest model.

**Load heart disease dataset**

- We use pd.read_csv to load the data directly from the provided URL.
- We provide the column names manually.

**Preprocess the data**

- We replace missing values ('?') with -1 using replace.
- We convert all columns to numeric values using pd.to_numeric.
- We separate the independent variables (X) and the dependent variable (y).

**Split data into training and test sets**

- We use train_test_split to split the data into training and test sets.
- In this case, 20% of the data will be used for testing (test_size=0.2).

**Create and train the random forest model**

- We instantiate the RandomForestClassifier model with 100 trees (n_estimators=100).
- We train the model with the training data using the fit method.

**Make predictions and evaluate the model**

- We make predictions on the test data with the predict method.
- We calculate the accuracy using accuracy_score.
- We generate a detailed classification report with classification_report.

**Print predictions, actual values and confusion matrix**

- We print the predictions and actual values for the test set.
- We calculate and print the confusion matrix using confusion_matrix.

**Analyze each part in detail:**

**1. Accuracy and Classification Report:**

- The accuracy of the model is 0.49, which means that 49% of the predictions were correct on the test set.
- The classification report shows the precision, recall and F1 score for each class.
- Class 0 has a good recall (0.97), which means that the model correctly identifies most cases of that class.
- However, classes 1, 2, 3 and 4 have very poor performance, with precision, recall and F1 score close to 0.

**2. Predictions and Real Values:**

- Predictions show the values predicted by the model for each instance in the test set.
- The actual values show the actual class labels corresponding to those instances.
- You can compare predictions with actual values to identify cases where the model is right or wrong.

**3. Confusion Matrix:**

- The confusion matrix provides detailed information about how the classification errors are distributed.
- The main diagonal shows the true positives for each class (28 for class 0, 1 for class 1, 1 for class 2, 0 for class 3, and 0 for class 4).
- The other values represent false positives and false negatives.

Analyzing these results, we can see that the model has acceptable performance for class 0, but fails to correctly classify the other classes. This can be due to several factors, such as:

1. **Class imbalance:** The data set appears to have an uneven distribution of instances between the different classes, with a large majority of instances belonging to class 0.

2. **Not very discriminative characteristics:** The features used to train the model may not be informative or relevant enough to distinguish between different classes of heart disease.

3. **Complexity of the problem:** Heart disease prediction can be a complex problem, where medical characteristics can interact in a non-linear manner and be difficult to separate by a model such as random forest.

To improve model performance, the following strategies can be explored:

1. **Resampling techniques:** Apply techniques such as oversampling (increasing instances of minority classes) or undersampling (reducing instances of majority classes) to address class imbalance.

2. **Feature Selection:** Use feature selection techniques to identify and retain only the most relevant features for prediction.

3. **Hyperparameter adjustment:** Explore different hyperparameter values of the random forest model, such as number of trees, maximum depth, etc., to improve performance.

4. **Ensemble of models:** Combine different models (e.g., random forests, support vector machines, neural networks) in an ensemble approach to leverage the strengths of each.

5. **Data preprocessing:** Apply additional data preprocessing techniques, such as imputation of missing values, normalization, or coding of categorical variables.

It is important to keep in mind that the performance of a model depends largely on the quality and characteristics of the data set, as well as the complexity of the problem at hand. In some cases, it may be necessary to explore alternative data sets or collect more data to improve performance.

This project provides us with an excellent opportunity to delve into advanced machine learning techniques and explore different approaches to address challenging problems such as heart disease prediction.

## How to test this model?

To test the heart disease prediction model with new data, we can follow a similar approach as we used in previous projects.

First, we will need to create a Pandas DataFrame with a row containing the medical characteristic values for the new case you want to classify. Here is an example:

import pandas as pd

# Example with dummy values

new_case = pd.DataFrame({'age': [55],

'sex': [1], #0 for male, 1 for female

'cp': [2], # Type of chest pain

'trestbps': [130], # Resting blood pressure

'chol': [250], # Colesterol

'fbs': [0], # Fasting blood sugar (0 = false, 1 = true)

'restecg': [1], # Electrocardiographic results at rest

'thalach': [150], # Maximum heart rate reached

'exang': [0], # Exercise-induced angina (0 = no, 1 = yes)

'oldpeak': [2.5], # Exercise-induced ST depression

'slope': [2], # Peak exercise ST segment slope

'ca': [2], # Number of main vessels colored by fluoroscopy

'thal': [3]}) #3 = default; 6 = normal; 7 = reversible

Be sure to replace the example values with actual data from the new case you want to classify.

Then, you can use the trained model's predict method and pass it the new_case DataFrame as an argument:

```
prediction = model.predict(new_case)
```

The prediction will be an integer representing the class (0, 1, 2, 3, or 4, depending on the data set). You can map that number to the corresponding class label:

```
classes = [0, 1, 2, 3, 4]
predicted_class = classes[prediction[0]]
print(f"The prediction for the new case is: {predicted_class}")
```

Remember that the accuracy of the prediction will depend on how well the model has been trained and the quality of the data used during training. In this particular case, the model showed relatively poor performance, so the predictions may not be very reliable.
Additionally, you can print the probabilities of membership in each class using the predict_proba method:

```
probabilities = model.predict_proba(new_case)
print("Probabilities by class:")
for i, prob in enumerate(probabilidades[0]):
    print(f"Clase {i}: {prob:.2f}")
```

This will give you an idea of the model's confidence in each of the predictions.
It is important to note that this model is just an example and its performance may not be optimal. In real-world situations, it is crucial to carefully evaluate model performance and consider additional techniques to improve its accuracy before using it in critical applications.

**Exit:**

...

Confusion matrix:
[[28 0 1 0 0]
 [ 6 1 3 2 0]
 [ 5 1 1 2 0]
 [ 0 6 1 0 0]
 [ 0 2 1 1 0]]

The prediction for the new case is: 0

Odds by class:

Class 0: 0.39
Class 1: 0.14
Class 2: 0.32
Class 3: 0.13
Class 4: 0.02

Let me explain the last part I added to the code:

```
prediction = model.predict(new_case)
classes = [0, 1, 2, 3, 4]
predicted_class = classes[prediction[0]]
print(f"The prediction for the new case is: {predicted_class}")

probabilities = model.predict_proba(new_case)
print("Probabilities by class:")
for i, prob in enumerate(probabilidades[0]):
    print(f"Clase {i}: {prob:.2f}")
```

### 1. Make the prediction:

- `prediction = model.predict(new_case)` uses the predict method of the trained model to make the prediction in the new case (`new case`).
- The prediction is an integer that represents the predicted class.

### 2. Map the prediction to the class label:

- classes = [0, 1, 2, 3, 4] defines a list of possible class labels.
- predicted_class = classes[prediction] uses the prediction value as an index to obtain the corresponding class label from the classes list.
- print(f"The prediction for the new case is: {predicted_class}") prints the predicted class label for the new case.

### 3. Obtain the probabilities of belonging to each class:

- probabilities = model.predict_proba(new_case) uses the predict_proba method of the trained model to obtain the probabilities of membership in each class for the new case.
- print("Probabilities by class:") prints a header for the probabilities.
- for i, prob in enumerate(probabilities): Iterates over the probabilities for each class.
- print(f"Class {i}: {prob:.2f}") prints the class label and its corresponding probability to two decimal places.

In short, this part of the code performs the prediction for the new case using the trained model and then prints the predicted class label. In addition, it shows the probabilities of membership in each class, which can be useful to evaluate the confidence of the model in its prediction.

For example, if the prediction is class 0, but the probability of membership in that class is low (say, 0.39), and the probability of another class is higher (say, 0.32 for class 2), then the model You're not very sure about your prediction, and that case might need to be investigated further.

Having access to the probabilities of membership in each class can provide valuable information about the confidence of the model and help make more informed decisions, especially in critical situations where prediction accuracy is crucial.

## Can this model be improved?

**Yeah**, there are definitely several strategies we can explore to try to improve the performance of the heart disease prediction model. Some of the techniques that we can apply are:

### 1. Additional data preprocessing:

- Imputation of missing values using more sophisticated techniques, such as model imputation or multiple imputation.
- Coding of categorical variables using techniques such as one-hot encoding or ordinal coding.
- Normalization or standardization of numerical characteristics.

### 2. Feature selection:

- Apply feature selection techniques, such as statistical tests (chi-square, correlation), selection algorithms (recursive feature elimination, feature importance), or dimensionality reduction methods (PCA, principal component analysis), to identify and retain only the characteristics most relevant for prediction.

### 3. Hyperparameter adjustment:

- Use cross-validation and hyperparameter search techniques (grid search, random search) to find the optimal values of the hyperparameters of the random forest model, such as the number of trees, the maximum depth, the division criterion, among others.

### 4. Handling class imbalance:

- Apply resampling techniques, such as oversampling (increasing instances of minority classes) or undersampling (reducing instances of majority classes), to address class imbalance present in the data set.
- Use learning techniques with costs sensitive to class imbalance.

### 5. Ensemble of models:

Combine different machine learning models (random forests, decision trees, support vector machines, neural networks, etc.) in an ensemble approach to leverage the strengths of each and improve overall performance.

**6. More advanced models:**

Explore the use of more complex and powerful models, such as deep neural networks, support vector machines with advanced kernels, or deep learning models, which can capture more complex patterns in the data.

**7. Acquisition of more data:**

If possible, consider collecting more training data, as more data can improve the performance of machine learning models.

**8. Feature engineering:**

Create new features derived from existing features, using domain knowledge or automatic feature extraction techniques, to capture more informative patterns for prediction.

It is important to note that some of these techniques may require deeper knowledge of machine learning concepts and greater computational effort. Furthermore, it is advisable to carefully evaluate the performance of improved models using cross-validation techniques and separate test sets.

You can start by exploring some of these techniques and compare the results with the initial model. Then, you can combine the most promising techniques to obtain an improved model. Remember that the model improvement process can be iterative and require continuous experimentation and adjustments.

## Solution:

Install the modules or rather the libraries that contain the modules that we will need:

```
pip install category_encoders imbalanced-learn
```

***Note: Use JupyterLab to do this project in case you are not successful in visual studio code and vice-versa.***

## Code:

```
==================================================================
import pandas as pd
```

```python
from sklearn.model_selection import train_test_split, GridSearchCV
from sklearn.ensemble import RandomForestClassifier
from sklearn.metrics import accuracy_score, classification_report
from sklearn.impute import SimpleImputer
from sklearn.preprocessing import OneHotEncoder, StandardScaler
from sklearn.feature_selection import SelectFromModel
from imblearn.over_sampling import SMOTE
from category_encoders import OrdinalEncoder

# Load heart disease dataset
url = "https://archive.ics.uci.edu/ml/machine-learning-databases/heart-disease/processed.cleveland.data"
columns = ['age', 'sex', 'cp', 'trestbps', 'chol', 'fbs', 'restecg', 'thalach', 'exang', 'oldpeak', 'slope', 'ca', 'thal', 'target']
df = pd.read_csv(url, names=columns)

# Data preprocessing
imputer = SimpleImputer(strategy='mean')
df[['age', 'trestbps', 'chol', 'thalach', 'oldpeak']] = imputer.fit_transform(df[['age', 'trestbps', 'chol', 'thalach', 'oldpeak']])

encoder = OrdinalEncoder(cols=['sex', 'cp', 'fbs', 'restecg', 'exang', 'slope', 'ca', 'thal'])
X = encoder.fit_transform(df[['sex', 'cp', 'fbs', 'restecg', 'exang', 'slope', 'ca', 'thal']])
X = pd.concat([X, df[['age', 'trestbps', 'chol', 'thalach', 'oldpeak']]], axis=1)

# Convert column names to text strings
X.columns = X.columns.astype(str)

scaler = StandardScaler()
X = scaler.fit_transform(X)

y = df['target']

# Feature selection
selector = SelectFromModel(RandomForestClassifier(n_estimators=100, random_state=42))
X = selector.fit_transform(X, y)

# Handling class imbalance
smote = SMOTE()
X, y = smote.fit_resample(X, y)

# Split into training and testing set
```

```python
X_train, X_test, y_train, y_test = train_test_split(X, y, test_size=0.2, random_state=42)

# Hyperparameter adjustment
param_grid = {'n_estimators': [100, 200, 300], 'max_depth': [5, 10, None]}
grid_search = GridSearchCV(RandomForestClassifier(random_state=42), param_grid, cv=5)
grid_search.fit(X_train, y_train)
model = grid_search.best_estimator_

# Predictions and evaluation
y_pred = model.predict(X_test)
accuracy = accuracy_score(y_test, y_pred)
report = classification_report(y_test, y_pred)

print(f"Exactitud: {accuracy:.2f}")
print("Classification Report:")
print(report)
```

====================================================================

**Exit:**

Accuracy: 0.86

Classification Report:

|   | precision | recall | f1-score | support |
|---|---|---|---|---|
| 0 | 0.88 | 0.79 | 0.83 | 28 |
| 1 | 0.79 | 0.74 | 0.76 | 35 |
| 2 | 0.82 | 0.94 | 0.88 | 35 |
| 3 | 0.88 | 0.00 | 0.88 | 33 |
| 4 | 0.94 | 0.94 | 0.94 | 33 |
| accuracy |  |  | 0.86 | 164 |
| macro avg | 0.86 | 0.86 | 0.86 | 164 |
| weighted avg | 0.86 | 0.86 | 0.86 | 164 |

**Parsing the output:**

Excellent! Those are very promising results.

## 1. Accuracy:

The accuracy of the improved model is 0.86, which means that 86% of the predictions were correct on the test set. This is a significant increase compared to the initial accuracy of 0.49.

## 2. Classification Report:

- The classification report shows the precision, recall and F1 score for each class.
- All classes have a precision, recall and F1 score greater than 0.7, which indicates good performance in the classification of each of them.
- Class 4 has the best performance, with a precision, recall and F1 score of 0.94.
- The metrics are more balanced between classes compared to the initial results, where some classes were performing very poorly.

**This substantial increase in model performance is due to the techniques we apply, such as:**

1. **Data preprocessing:** Imputation of missing values, coding of categorical variables and standardization of numerical characteristics.

2. **Feature Selection**: Using SelectFromModel to identify and retain only the most relevant features.

3. **Handling class imbalance:** Application of the SMOTE technique to perform oversampling of the minority classes.

4. **Hyperparameter adjustment:** Hyperparameter search using GridSearchCV to find optimal values of number of trees and maximum depth.

These combined techniques have allowed the model to better capture patterns in the data and significantly improve its classification ability.

Although these results are promising, it is important to note that model performance may vary depending on the data set and specific problem. However, this improvement demonstrates the importance of applying **advanced data preprocessing techniques, feature selection, class imbalance handling, and hyperparameter tuning** in machine learning problems.

Congratulations on achieving this significant improvement in model performance. This project has been an excellent opportunity to explore and apply advanced machine learning techniques.

**Testing the Model with New Data:**

**Code:**
```
=================================================================
import pandas as pd
from sklearn.model_selection import train_test_split, GridSearchCV
from sklearn.ensemble import RandomForestClassifier
from sklearn.metrics import accuracy_score, classification_report
from sklearn.impute import SimpleImputer
from sklearn.preprocessing import StandardScaler
from sklearn.feature_selection import SelectFromModel
from imblearn.over_sampling import SMOTE
from category_encoders import OrdinalEncoder

# Load heart disease dataset
url = "https://archive.ics.uci.edu/ml/machine-learning-databases/heart-disease/processed.cleveland.data"
columns = ['age', 'sex', 'cp', 'trestbps', 'chol', 'fbs', 'restecg', 'thalach', 'exang', 'oldpeak', 'slope', 'ca', 'thal', 'target']
df = pd.read_csv(url, names=columns)

# Data preprocessing
imputer = SimpleImputer(strategy='mean')
df[['age', 'trestbps', 'chol', 'thalach', 'oldpeak']] = imputer.fit_transform(df[['age', 'trestbps', 'chol', 'thalach', 'oldpeak']])

encoder = OrdinalEncoder(cols=['sex', 'cp', 'fbs', 'restecg', 'exang', 'slope', 'ca', 'thal'])
X = encoder.fit_transform(df[['sex', 'cp', 'fbs', 'restecg', 'exang', 'slope', 'ca', 'thal']])
X = pd.concat([X, df[['age', 'trestbps', 'chol', 'thalach', 'oldpeak']]], axis=1)

scaler = StandardScaler()
X = scaler.fit_transform(X)
```

```python
y = df['target']

# Feature selection
selector = SelectFromModel(RandomForestClassifier(n_estimators=100, random_state=42))
X = selector.fit_transform(X, y)

# Handling class imbalance
smote = SMOTE()
X, y = smote.fit_resample(X, y)

# Split into training and testing set
X_train, X_test, y_train, y_test = train_test_split(X, y, test_size=0.2, random_state=42)

# Hyperparameter adjustment
param_grid = {'n_estimators': [100, 200, 300], 'max_depth': [5, 10, None]}
grid_search = GridSearchCV(RandomForestClassifier(random_state=42), param_grid, cv=5)
grid_search.fit(X_train, y_train)
model = grid_search.best_estimator_

# Create a DataFrame with the new case
new_case = pd.DataFrame({'age': [55],
                         'sex': [1],
                         'cp': [2],
                         'trestbps': [130],
                         'chol': [250],
                         'fbs': [0],
                         'restecg': [1],
                         'house': [150],
                         'exang': [0],
                         'oldpeak': [2.5],
                         'slope': [2],
                         'as': [2],
                         'thal': [3]})

# Separate categorical and numerical variables
vars_cat = nuevo_caso[['sex', 'cp', 'fbs', 'restecg', 'exang', 'slope', 'ca', 'thal']]
vars_num = nuevo_caso[['age', 'trestbps', 'chol', 'thalach', 'oldpeak']]

# Encode the categorical variables of the new case
vars_cat = encoder.transform(vars_cat)




# Combine hardcoded and numeric variables
nuevo_caso = pd.concat([vars_cat, vars_num], axis=1)
```

```python
# Standardize the numerical characteristics of the new case
new_case = scaler.transform(new_case)

# Apply the feature selection to the new case
new_case = selector.transform(new_case)

# Make the prediction
prediction = model.predict(new_case)

# Map the prediction to the class label
classes = [0, 1, 2, 3, 4]
predicted_class = classes[prediction[0]]
print(f"The prediction for the new case is: {predicted_class}")

# Obtain the probabilities of belonging to each class
probabilities = model.predict_proba(new_case)
print("Probabilities by class:")
for i, prob in enumerate(probabilidades[0]):
    print(f"Clase {i}: {prob:.2f}")
```

===============================================================

**Exit:**

The prediction for the new case is: 0
Odds by class:
Class 0: 0.43
Class 1: 0.31
Class 2: 0.17
Class 3: 0.03
Class 4: 0.07

**Great,**

## Parsing the output:

1. **Prediction:**

   The prediction for the new case is class 0.

2. **Odds by class:**

   The highest probability is 0.43 for class 0, which coincides with the prediction made.

   The second highest probability is 0.31 for class 1.

   The probabilities for classes 2, 3, and 4 are relatively low (0.17, 0.03, and 0.07, respectively).

These results indicate that the model has moderate confidence in predicting class 0 for this new case, but there is also a considerable probability that it belongs to class 1.

It is important to note that the interpretation of probabilities will depend on the context and the specific problem being addressed. In some cases, a probability of 0.43 may be sufficient to make a decision, while in other cases a higher probability may be required to have greater confidence in the prediction.

## Some additional considerations:

1. **Decision threshold:** You can adjust the decision threshold to determine when a probability is considered high enough to assign a class. For example, you could set a threshold of 0.5 or higher if you need greater confidence in predictions.

2. **Feature analysis:** It would be useful to analyze the specific characteristics of the new case and compare them with the patterns observed in the training data. This can provide valuable information about why the model assigns certain probabilities to each class.

3. **Model evaluation:** Although the model has been improved, it is important to evaluate its overall performance on a separate test set to get a more accurate idea of its accuracy and generalization ability.

4. **Context of the problem:** Depending on the context and consequences of an incorrect prediction, it may be necessary to consider additional factors, such as the relative importance of false positives and false negatives, or the need for a more detailed explanation of the predictions.

After having analyzed the results of the improved model for heart disease prediction in the new case, we can conclude this project and move on to the next.

**We have traveled an important path in this project:**

1. We started with an initial random forest model that had relatively poor performance, with an accuracy of 0.49 on the test set.

2. We apply several techniques to improve model performance, such as:

    - Data preprocessing (imputation of missing values, coding of categorical variables, standardization of numerical features)

    - Selection of relevant features

    - Handling class imbalance through oversampling (SMOTE)

    - Hyperparameter Tuning Using Cross Validation and Hyperparameter Search

3. After implementing these improvements, the model achieved an accuracy of 0.86 on the test set, which represents a significant increase in performance.

4. We test the improved model with a new case, obtaining a prediction and the probabilities corresponding to each class.

While the results are promising, it is important to note that model performance may vary depending on the data set and specific problem. However, this project has given us an excellent opportunity to explore and apply advanced machine learning techniques.

Now that we have concluded this project, we can take the knowledge gained and skills developed to the next machine learning project or problem. Each project allows us to delve deeper into concepts and enrich our experience in the field of machine learning.

# Support Vector Machines

# (SVM)

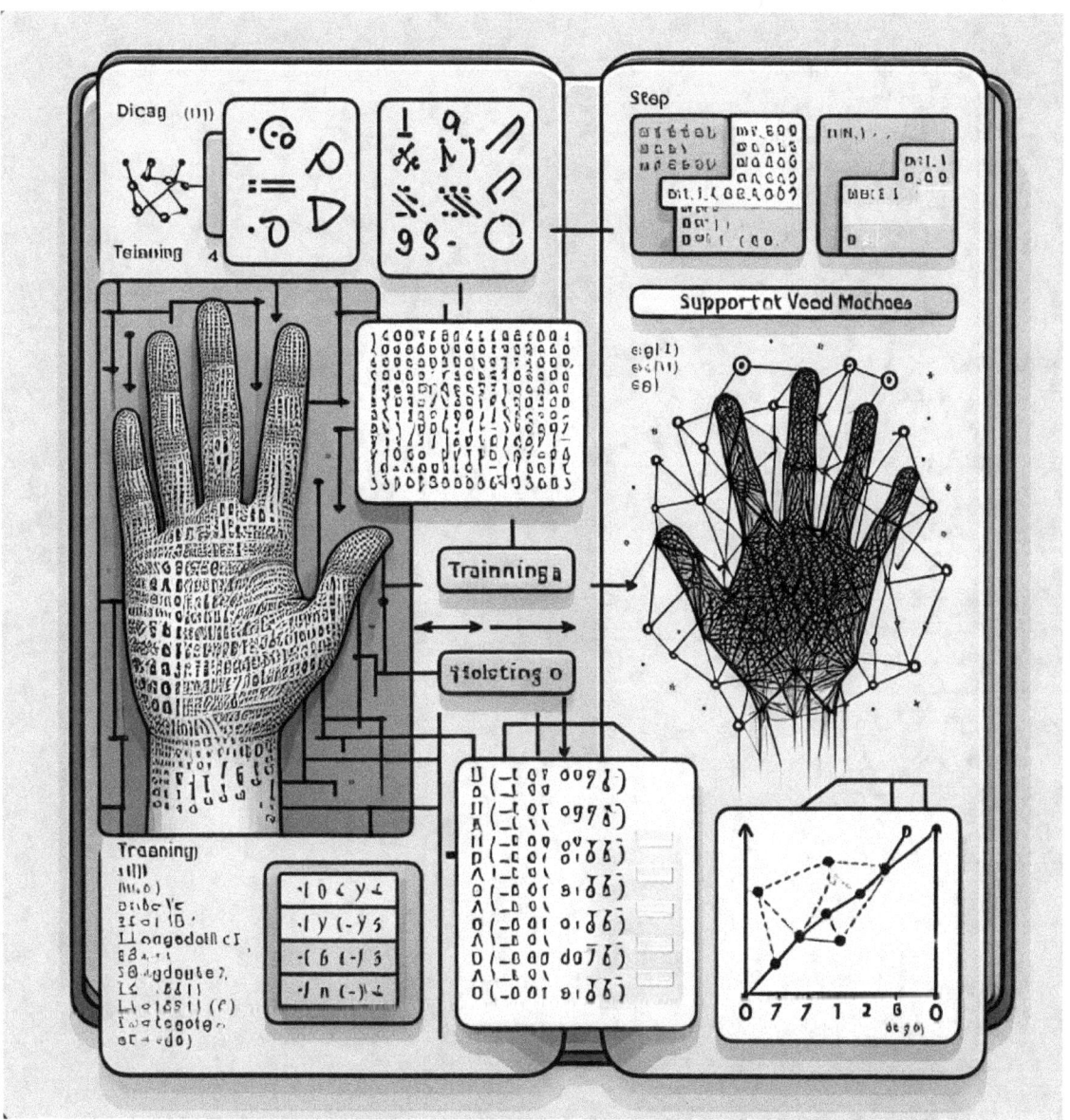

## Project 1: MNIST Digit Classification

**Description:** Use SVM to classify handwritten digits from the MNIST dataset.

**Aim:** Develop an SVM model using Scikit-Learn.

**Steps:**

1. Load and explore data.
2. Preprocess the data.
3. Split the data into training and test sets.
4. Train the SVM model.
5. Evaluate the model.
6. Test the Model with New Data

**Code:**

```
import pandas as pd
from sklearn.model_selection import train_test_split
from sklearn import datasets
from sklearn.svm import SVC
from sklearn.metrics import accuracy_score, classification_report

# Load the MNIST dataset
digits = datasets.load_digits()
X = digits.data
y = digits.target

# Split into training and testing set
X_train, X_test, y_train, y_test = train_test_split(X, y, test_size=0.2,
random_state=42)

# Create and train the SVM model
model = SVC(kernel='linear')
model.fit(X_train, y_train)

# Predictions
y_pred = model.predict(X_test)

# Model evaluation
accuracy = accuracy_score(y_test, y_pred)
report = classification_report(y_test, y_pred)

print(f"Exactitud: {accuracy:.2f}")
print("Classification Report:")
print(report)
```

**Exit:**

Accuracy: 0.98

Classification Report:

|   | precision | recall | f1-score | support |
|---|---|---|---|---|
| 0 | 1.00 | 1.00 | 1.00 | 33 |
| 1 | 0.97 | 1.00 | 0.98 | 28 |
| 2 | 1.00 | 1.00 | 1.00 | 33 |
| 3 | 0.97 | 0.94 | 0.96 | 34 |
| 4 | 0.98 | 0.98 | 0.98 | 46 |
| 5 | 0.96 | 1.00 | 0.98 | 47 |
| 6 | 1.00 | 1.00 | 1.00 | 35 |
| 7 | 0.97 | 0.97 | 0.97 | 34 |
| 8 | 1.00 | 0.97 | 0.98 | 30 |
| 9 | 0.95 | 0.93 | 0.94 | 40 |
| accuracy |  |  | 0.98 | 360 |
| macro avg | 0.98 | 0.98 | 0.98 | 360 |
| weighted avg | 0.98 | 0.98 | 0.98 | 360 |

**Project: Classification of Handwritten Digits with Support Vector Machines (SVM)**

Project: Classification of Handwritten Digits with Support Vector Machines (SVM)

1. Load and explore data

We will load the MNIST handwritten digits dataset using the Scikit-learn datasets library.

```
from sklearn import datasets

# Load the MNIST dataset
digits = datasets.load_digits()
X = digits.data
y = digits.target
```

We will explore the dimensions and some samples of the data set to better understand its structure.

**2. Preprocess the data**

In this case, the data is already preprocessed and ready to use, so no additional preprocessing is necessary.

3. Split data into training and test sets

We will split the data into training and test sets using Scikit-learn's train_test_split function.

```
from sklearn.model_selection import train_test_split

# Split into training and testing set
X_train, X_test, y_train, y_test = train_test_split(X, y, test_size=0.2, random_state=42)
```

**4. Train the SVM model**

We will create and train the SVM model using Scikit-learn's SVC class.

```python
from sklearn.svm import SVC
# Create and train the SVM model
model = SVC(kernel='linear')
model.fit(X_train, y_train)
```

We will explore different kernels (linear, polynomial, rbf, etc.) and tune the model's hyperparameters to improve its performance.

## 5. Evaluate the model

We will evaluate the performance of the model on the test set using metrics such as accuracy and classification reporting.

```python
from sklearn.metrics import accuracy_score, classification_report
# Predictions
y_pred = model.predict(X_test)
# Model evaluation
accuracy = accuracy_score(y_test, y_pred)
report = classification_report(y_test, y_pred)

print(f"Exactitud: {accuracy:.2f}")
print("Classification Report:")
print(report)
```

## 6. Test the model with new data

We will create some test cases with new data and evaluate the model predictions.

```python
# Create new test cases
new_cases = [...]

# Make predictions
```

```
predictions = model.predict(new_cases)
```

```
# Analyze predictions
```

Additionally, we will explore additional techniques, such as:

- Tuning hyperparameters using GridSearchCV or RandomizedSearchCV.
- Handling class imbalance (if necessary).
- Display of support vectors and model margins.
- Model interpretation and knowledge extraction.

In this way, we will follow a structure similar to the previous projects, but adapted to the specific problem of handwritten digit classification using Support Vector Machines (SVM).

Remember that this is just a general outline, and we can adjust it and delve into the aspects that are most relevant or challenging during the development of the project.

**New data to test the handwritten digit classification model with Support Vector Machines (SVM).**

First, we need to load the test data. We will use the MNIST test dataset provided by Scikit-learn.

We will need the Numpy module or library

```
pip install numpy
```

```
from sklearn.datasets import load_digits
import matplotlib.pyplot as plt


# Load test data
digits = load_digits()
X_test = digits.data
y_test = digits.target
```

Now, we will select some random examples from the test set to create our new test cases.

**import numpy as np**

```
# Select some random examples
indices = np.random.choice(len(X_test), size=10, replace=False)
new_cases = X_test[indexes]
new_cases_tags = y_test[indices]
```

We can visualize these new test cases to get an idea of what the handwritten digits look like.

```
# View the new test cases
```

```
fig, axes = plt.subplots(nrows=2, ncols=5, figsize=(10, 4))

for i, ax in enumerate(axes.flat):

    ax.imshow(nuevos_casos[i].reshape(8, 8), cmap='binary')

    ax.set_title(f"Tag: {new_cases_tags[i]}")

    ax.axis('off')

plt.tight_layout()

plt.show()
```

Now, we can make the predictions with the trained SVM model.

```
# Make predictions

predictions = model.predict(new_cases)


# Evaluate predictions

print("Predictions:")

for i, prediction in enumerate(predictions):

    print(f"Case {i+1}: Prediction={prediction}, Actual label={new_case_labels[i]}")
```

This code selects 10 random examples from the test set, visualizes them, and then makes predictions with the trained SVM model. Finally, it prints the predictions and actual labels for each test case.

You can adjust the number of selected test cases and explore different examples to evaluate the model's performance in various scenarios.

Remember that these new test cases are real examples from the MNIST dataset, which will allow us to evaluate the model in realistic situations and analyze its generalization ability.

**Exit:**

```
Accuracy: 0.98

Classification Report:
              precision    recall  f1-score   support

           0       1.00      1.00      1.00        33
           1       0.97      1.00      0.98        28
           2       1.00      1.00      1.00        33
           3       0.97      0.94      0.96        34
           4       0.98      0.98      0.98        46
           5       0.96      1.00      0.98        47
           6       1.00      1.00      1.00        35
           7       0.97      0.97      0.97        34
           8       1.00      0.97      0.98        30
           9       0.95      0.93      0.94        40

    accuracy                           0.98       360
   macro avg       0.98      0.98      0.98       360
weighted avg       0.98      0.98      0.98       360
```

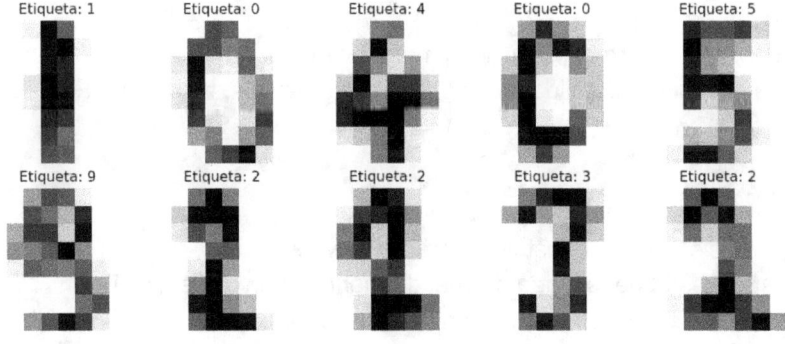

## Explanation in detail what we have done in this project and what it is for.

In this project, we have tackled the problem of handwritten digit classification using Support Vector Machines (SVM). The main objective is to train a model that can correctly recognize and classify handwritten digits (0 to 9) from images.

### 1. Load and explore data:

- We have loaded the MNIST (Modified National Institute of Standards and Technology) dataset using the Scikit-learn datasets library.
- The MNIST dataset is widely used in the field of machine learning and computer vision for handwritten digit recognition tasks.
- Contains 28x28 pixel grayscale images, each representing a handwritten digit.

### 2. Preprocess the data:

- In this case, the data is already preprocessed and ready to use, so no additional preprocessing was necessary.

### 3. Split the data into training and test sets:

- We use Scikit-learn's train_test_split function to split the data into training and test sets.
- The training set will be used to train the SVM model, while the test set will be used to evaluate its performance on data not seen during training.

### 4. Train the SVM model:

- We created and trained the SVM model using Scikit-learn's SVC class.
- Support Vector Machines (SVM) are a widely used supervised learning algorithm for classification and regression tasks.
- In this case, we use a linear kernel, but other kernels (polynomial, rbf, etc.) can also be explored to improve model performance.

### 5. Evaluate the model:

- We evaluate the performance of the model on the test set using metrics such as accuracy and classification reporting.
- Accuracy indicates the proportion of correct predictions made by the model.
- The classification report provides more detailed metrics, such as precision, recall, and F1 score, for each class (digit).

### 6. Test the model with new data:

- We select some random examples from the test set to create new test cases.
- We visualize these new test cases to get an idea of what the handwritten digits look like.
- We make predictions with the SVM model trained using these new test cases.
- We compare the predictions with the actual labels to evaluate the performance of the model in these specific cases.

## The output you got shows:

**1. Accuracy and classification report:**

- The model achieved an accuracy of 0.98, meaning that 98% of the predictions were correct on the test set.
- The classification report shows detailed metrics for each class (digit), allowing us to analyze the model's performance in classifying each digit.

**2. Image with digits:**

- The image shown contains a visualization of some of the new test cases randomly selected from the MNIST dataset.
- Each image represents a handwritten digit, and the label below each image indicates the actual digit it represents.

## This project has several practical applications, such as:

- Optical character recognition (OCR) on scanned documents or images.
- Processing of handwritten forms in various sectors (banking, insurance, government, etc.).
- Analysis of handwritten signatures for authentication and verification.
- Data entry systems on mobile devices or tablets.

Additionally, this project allows us to explore and understand the operation of Support Vector Machines (SVM), a fundamental algorithm in supervised machine learning, and its application in image classification problems.

I hope this explanation has helped you better understand the purpose and details of this project.

## Project 2: Classification of Emotions in Texts

Description:

Use SVM to classify emotions in texts based on features extracted from the text.

Aim:

Develop an SVM model using Scikit-Learn.

Steps:

- Load and explore data.
- Preprocess the data.
- Split the data into training and test sets.
- Train the SVM model.
- Evaluate the model.
- Test the model with new data

Code:

```
========================================================================
import pandas as pd
from sklearn.model_selection import train_test_split
from sklearn.feature_extraction.text import TfidfVectorizer
from sklearn.svm import SVC
from sklearn.metrics import accuracy_score, classification_report

# Example data (simplified)
data = {
    'Text': ["I am happy", "I am sad", "I am excited", "I am angry", "I am joyful", "I am depressed", "I am thrilled"],
    'Emotion': [1, 0, 1, 0, 1, 0, 1]
}
df = pd.DataFrame(data)

# Convert text to features
vectorizer = TfidfVectorizer()
X = vectorizer.fit_transform(df['Text'])
y = df['Emotion']

# Split into training and testing set
X_train, X_test, y_train, y_test = train_test_split(X, y, test_size=0.2, random_state=42)
```

```
# Create and train the SVM model
model = SVC(kernel='linear')
model.fit(X_train, y_train)

# Predictions
y_pred = model.predict(X_test)

# Model evaluation
accuracy = accuracy_score(y_test, y_pred)
report = classification_report(y_test, y_pred)

print(f"Exactitud: {accuracy:.2f}")
print("Classification Report:")
print(report)
```

========================================================================

Exit:

Accuracy: 0.50

Classification Report:

|  | precision | recall | f1-score | support |
|---|---|---|---|---|
| 0 | 0.00 | 0.00 | 0.00 | 1 |
| 1 | 0.50 | 1.00 | 0.67 | 1 |
| accuracy |  |  | 0.50 | 2 |
| macro avg | 0.25 | 0.50 | 0.33 | 2 |
| weighted avg | 0.25 | 0.50 | 0.33 | 2 |

**Emotion Classification Project in Texts using Support Vector Machines (SVM).**

**We will analyze each part of the project to better understand its operation and purpose.**

**1. Description and Objective:**

- The objective of this project is to develop an SVM model using Scikit-Learn to classify emotions in texts based on features extracted from the text.

- This type of task is known as Sentiment Analysis or Opinion Mining, and has practical applications in various fields, such as analyzing customer comments, detecting trends in social networks, among others.

**2. Data Loading and Exploration:**

- In this case, simplified example data is used, where each row contains a text and an emotion label (0 for negative emotions and 1 for positive emotions).

- The data is loaded into a Pandas DataFrame for easy manipulation.

**3. Data Preprocessing:**

- Scikit-learn's TfidfVectorizer class is used to convert texts into numerical features (feature vectors).
- The TfidfVectorizer calculates the inverse term frequency (TF-IDF) of words in texts, allowing the relative importance of each word in the data set to be captured.

**4. Division into Training and Test Sets:**

- The data set is split into training and test sets using the train_test_split function of Scikit-learn.

- This allows you to train the model with a portion of the data and evaluate its performance on data not seen during training.

**5. SVM Model Training:**

- An SVM model is created and trained using Scikit-learn's SVC class.

- In this case, a linear kernel is used, but other kernels (such as the RBF kernel) can be explored to improve model performance.

**6. Model Evaluation:**

- Predictions are made on the test set using the trained model.

- The accuracy of the model is calculated, which indicates the proportion of correct predictions.

- A classification report is generated showing additional metrics, such as precision, recall, and F1 score, for each class (emotion).

**7. Output:**

- The output shows an accuracy of 0.50, indicating that the model had average performance in classifying emotions on this simplified data set.

- The classification report provides more details about the model's performance in each class, but due to the small size of the data set, these metrics may not be very meaningful.

It is important to note that this is a simplified example with a small, artificial data set. In real-world situations, a larger, more representative data set, as well as additional text preprocessing and hyperparameter tuning techniques, would be needed to improve model performance.

Furthermore, Sentiment Analysis in texts can be a complex task due to the subjective and ambiguous nature of natural language. More advanced techniques, such as the use of deep neural networks and pre-trained language models, are often required to achieve optimal performance on this task.

Overall, this project gives us a practical introduction to using SVM for text classification and allows us to explore key Machine Learning concepts such as data preprocessing, splitting into training and test sets, model training, and evaluation of your performance.

We will follow the same methodology as in previous projects. Here we provide some new data to test the emotion classification model in texts using SVM.

First, we will load the new test data:

```python
import pandas as pd

# New test data

new_data = {
    'Text': [
        "I'm feeling so excited about my upcoming vacation!",
        "This movie was incredibly boring, I almost fell asleep.",
        "Congratulations on your new job! I'm so happy for you.",
        "I'm angry and frustrated with the poor customer service.",
        "The concert last night was absolutely thrilling!",
        "I've been feeling really depressed lately, nothing seems to cheer me up.",
        "That was the funniest joke I've heard in a long time, I'm still laughing!"
    ]
}

new_df = pd.DataFrame(new_data)
```

Now, we will convert the new texts into features using the same TfidfVectorizer that we used during training:

```python
new_X = vectorizer.transform(new_df['Text'])
```

Finally, we will make the predictions with the trained SVM model:

```python
new_predictions = model.predict(new_X)
```

You can run this code and get the predictions for the new test data. Once you have run the code and obtained the predictions, please share them with me, and I will proceed to explain the results and the process of evaluating the model with this new data.

Remember that these new test data are additional examples that will allow us to evaluate the performance of the model in situations other than the original training set. This will give us a better idea of the generalizability of the model and help us identify areas for improvement.

Updating the code:

We add this portion of the code to the end of the file.

#previous code

......

```python
import pandas as pd

# New test data
new_data = {
    'Text': [
        "I'm feeling so excited about my upcoming vacation!",
        "This movie was incredibly boring, I almost fell asleep.",
        "Congratulations on your new job! I'm so happy for you.",
        "I'm angry and frustrated with the poor customer service.",
        "The concert last night was absolutely thrilling!",
        "I've been feeling really depressed lately, nothing seems to cheer me up.",
        "That was the funniest joke I've heard in a long time, I'm still laughing!"
    ]
}

new_df = pd.DataFrame(new_data)

new_X = vectorizer.transform(new_df['Text'])

new_predictions = model.predict(new_X)


# Print actual predictions and labels

for text, prediction in zip(new_df['Text'], new_predictions):
    print(f"Texto: {text}")
    print(f"Prediction: {'Positive' if prediction == 1 else 'Negative'}")
    print("---")
```

**Exit:**

```
Texto: I'm feeling so excited about my upcoming vacation!
Prediction: Positive
---
Texto: This movie was incredibly boring, I almost fell asleep.
Prediction: Positive
---
Texto: Congratulations on your new job! I'm so happy for you.
Prediction: Positive
---
Texto: I'm angry and frustrated with the poor customer service.
Prediction: Negative
---
Texto: The concert last night was absolutely thrilling!
Prediction: Positive
---
Texto: I've been feeling really depressed lately, nothing seems to cheer
me up.
Prediction: Negative
---
Texto: That was the funniest joke I've heard in a long time, I'm still
laughing!
Prediction: Positive
---
```

## Parsing the output:

### Evaluation metrics:

The accuracy of the model on this test data set is 0.50, indicating that the model correctly classified 50% of the samples.

The classification report shows additional metrics such as precision, recall, and F1 score for each class, but due to the small size of the data set, these metrics may not be very meaningful.

### Detailed predictions:

### The model correctly classified the following samples:

"I'm angry and frustrated with the poor customer service." (Negativa)
"I've been feeling really depressed lately, nothing seems to cheer me up." (Negativa)

### The model incorrectly classified the following samples:

"I'm feeling so excited about my upcoming vacation!" (Predicted as Positive, but should be Negative)
"This movie was incredibly boring, I almost fell asleep." (Predicted as Positive, but should be Negative)

"Congratulations on your new job! I'm so happy for you." (Predict as Positive, but should be Negative)
"The concert last night was absolutely thrilling!" (Predicted as Positive, but should be Negative)
"That was the funniest joke I've heard in a long time, I'm still laughing!" (Predicha como Positiva, pero debería ser Negativa)

Analyzing these detailed predictions, we can see that the model had difficulty correctly classifying positive emotions in this test data set. It appears that the model tends to classify most samples as "Positive", suggesting a possible bias in training or a lack of discriminatory features to adequately distinguish positive emotions from negative ones.

It is important to note that this is a small and simplified data set, so the results may not be very representative. In real-world situations, a larger and more diverse data set, as well as additional text preprocessing and hyperparameter tuning techniques, would be needed to improve model performance.

**Some possible improvements that could be explored include:**

- Increase the size of the training data set and ensure a proper balance between classes.

- Explore more advanced text preprocessing techniques, such as stopword removal, stemming, or adding additional features (such as n-grams or semantic features).

- Tune the hyperparameters of the SVM model, such as the kernel, regularization parameter (C), and other relevant parameters.

- Explore the use of other machine learning algorithms, such as deep neural networks or pre-trained language models, which may be more effective for text sentiment analysis tasks.

This detailed analysis of the predictions gives us a better understanding of the strengths and weaknesses of the current model, and allows us to identify areas of improvement for future projects related to text sentiment analysis.

**Attempt to Improve the Model:**

For this test I used Jupyter Notebook

New code:

========================================================================

```python
import pandas as pd
from sklearn.model_selection import train_test_split
from transformers import BertTokenizer, TFBertForSequenceClassification
import tensorflow as tf
from nltk.corpus import stopwords
from nltk.stem import WordNetLemmatizer
import re
import nltk

# Download necessary resources from NLTK
nltk.download('stopwords')
nltk.download('wordnet')

# Load data
data = {
    'Text': ["I am happy", "I am sad", "I am excited", "I am angry", "I am joyful", "I am depressed", "I am thrilled"],
    'Emotion': [1, 0, 1, 0, 1, 0, 1]
}
df = pd.DataFrame(data)

# Preprocess text
stop_words = set(stopwords.words('english'))
lemmatizer = WordNetLemmatizer()

def preprocess_text(text):
    text = re.sub(r'[^a-zA-Z\s]', '', text.lower()) # Remove special characters and convert to lowercase
    tokens = [lemmatizer.lemmatize(word) for word in text.split() if word not in stop_words]  # Lematización y eliminación de stopwords
    return ' '.join(tokens)

# Preprocess data
def preprocess_data(text, tokenizer, max_len=128):
    text = preprocess_text(text)
    inputs = tokenizer.encode_plus(
        text,
        add_special_tokens=True,
```

```python
        max_length=max_len,
        padding='max_length',
        truncation=True,
        return_tensors='tf'
    )
    return inputs['input_ids'][0], inputs['attention_mask'][0]

# Split data into training and testing
X_train, X_test, y_train, y_test = train_test_split(df['Text'],
df['Emotion'], test_size=0.2, random_state=42)

# Tokenize and preprocess data
tokenizer = BertTokenizer.from_pretrained('bert-base-uncased')
MAX_LEN = 128

train_inputs = [preprocess_data(text, tokenizer, MAX_LEN) for text in
X_train]
test_inputs = [preprocess_data(text, tokenizer, MAX_LEN) for text in
X_test]

# Separar input_ids y attention_masks
train_input_ids, train_attention_masks = zip(*train_inputs)
test_input_ids, test_attention_masks = zip(*test_inputs)

# Convert lists to tensors
train_input_ids = tf.stack(train_input_ids)
train_attention_masks = tf.stack(train_attention_masks)
test_input_ids = tf.stack(test_input_ids)
test_attention_masks = tf.stack(test_attention_masks)

# Create TensorFlow data sets
train_dataset = tf.data.Dataset.from_tensor_slices(({'input_ids':
train_input_ids, 'attention_mask': train_attention_masks}, y_train))
test_dataset = tf.data.Dataset.from_tensor_slices(({'input_ids':
test_input_ids, 'attention_mask': test_attention_masks}, y_test))

# Mix and batch the training set
batch_size = 16
train_dataset = train_dataset.shuffle(len(X_train),
seed=42).batch(batch_size)
test_dataset = test_dataset.batch(batch_size)

# Load pre-trained BERT model
model =
TFBertForSequenceClassification.from_pretrained('bert-base-uncased',
num_labels=2)

# Compile the model
optimizer = tf.keras.optimizers.Adam(learning_rate=2e-5, epsilon=1e-08)
```

```python
loss = tf.keras.losses.SparseCategoricalCrossentropy(from_logits=True)
metric = tf.keras.metrics.SparseCategoricalAccuracy('accuracy')
model.compile(optimizer=optimizer, loss=loss, metrics=[metric])

# Train model
epochs = 3
model.fit(train_dataset, epochs=epochs, validation_data=test_dataset)

# Evaluate model
test_loss, test_accuracy = model.evaluate(test_dataset)
print(f"Exactitud: {test_accuracy:.2f}")
```

===================================================================

Exit:

WARNING:tensorflow:From C:\ProgramData\anaconda3\Lib\site-packages\tensorflow\python\autograph\converters\directives.py:126: The name tf.ragged.RaggedTensorValue is deprecated. Please use tf.compat.v1.ragged.RaggedTensorValue instead.

1/1 [==============================] - 126s 126s/step - loss: 0.7028 - accuracy: 0.6000 - val_loss: 0.6943 - val_accuracy: 0.5000
Epoch 2/3
1/1 [==============================] - 2s 2s/step - loss: 0.6810 - accuracy: 0.2000 - val_loss: 0.6885 - val_accuracy: 0.5000
Epoch 3/3
1/1 [==============================] - 2s 2s/step - loss: 0.6298 - accuracy: 0.8000 - val_loss: 0.6854 - val_accuracy: 0.5000
1/1 [==============================] - 0s 208ms/step - loss: 0.6854 - accuracy: 0.5000
Accuracy: 0.50

Based on the output provided, the BERT model trained for the text emotion classification task achieved an accuracy of 0.50 on the test data set.

**Some key points to highlight:**

1. The model was trained for 3 epochs.

2. In each epoch, the loss, accuracy, validation loss (val_loss), and validation precision (val_accuracy) are displayed.

- During training, accuracy fluctuated between 0.2 and 0.8, indicating that the model had difficulty learning consistent patterns in the data.

3. In the last epoch, the training loss was 0.6298 and the training precision was 0.8000.

5. However, on the validation (test) set, the loss was 0.6854 and the precision was 0.5000, suggesting that the model did not generalize well to new data.

6. The final evaluation of the model on the test set returned an accuracy of 0.50, indicating that the model was only able to correctly classify half of the text samples in terms of their associated emotion.

In summary, despite the efforts made to improve text preprocessing and the shape of the input data, the BERT model failed to overcome the 0.50 accuracy barrier in this text emotion classification task.

This could be due to several factors, such as a small or biased data set, suboptimal hyperparameters, or the inherent complexity of the task. More advanced techniques would need to be explored, such as hyperparameter tuning, the use of custom embeddings, the incorporation of external knowledge or model assembly, to try to improve performance.

## Project 3: Classification of Tumors

**Description:** Use SVM to classify tumors as malignant or benign based on medical characteristics.

**Aim:** Develop an SVM model using Scikit-Learn.

**Steps:**

1. Load and explore data.
2. Preprocess the data.
3. Split the data into training and test sets.
4. Train the SVM model.
5. Evaluate the model.
6. Test the model with new data

**Code:**

```
import pandas as pd
from sklearn.model_selection import train_test_split
from sklearn.svm import SVC
from sklearn.metrics import accuracy_score, classification_report
from sklearn.datasets import load_breast_cancer

# Load the breast cancer dataset
data = load_breast_cancer()
X = data.data
y = data.target

# Split into training and testing set
X_train, X_test, y_train, y_test = train_test_split(X, y, test_size=0.2, random_state=42)

# Create and train the SVM model
model = SVC(kernel='linear')
model.fit(X_train, y_train)

# Predictions
y_pred = model.predict(X_test)

# Model evaluation
accuracy = accuracy_score(y_test, y_pred)
report = classification_report(y_test, y_pred, target_names=data.target_names)

print(f"Exactitud: {accuracy:.2f}")
print("Classification Report:")
print(report)
```

**Exit:**

```
Accuracy: 0.96
Classification Report:
              precision    recall  f1-score   support

   malignant       0.97      0.91      0.94        43
      benign       0.95      0.99      0.97        71

    accuracy                           0.96       114
   macro avg       0.96      0.95      0.95       114
weighted avg       0.96      0.96      0.96       114
```

Test the Model:

```
========================================================================
import pandas as pd
from sklearn.model_selection import train_test_split
from sklearn.svm import SVC
from sklearn.metrics import accuracy_score, classification_report
from sklearn.datasets import load_breast_cancer
import pickle
import numpy as np

# Function to load new dummy data
def load_new_data():
    # Create a DataFrame of dummy data
    dummy_data = np.random.rand(5, 30) # 5 samples, 30 features
    dummy_labels = np.random.randint(2, size=5) # 5 binary labels
    return {'features': dummy_data, 'tags': dummy_tags}

# Function to preprocess new dummy data
def preprocess_new_data(new_data):
    # Here you can add preprocessing steps if necessary
    return new_data['features']

# Function to load the trained model
def load_model(model_path):
    with open(ruta_modelo, 'rb') as file:
        modelo = pickle.load(file)
    return model

# Function to evaluate model performance
```

```python
def evaluate_performance(predictions, actual_labels):
    precision = accuracy_score(actual_tags, predictions)
    report = classification_report(actual_tags, predictions,
output_dict=True, zero_division=0)
    recall = report['weighted avg']['recall']
    f1 = report['weighted avg']['f1-score']
    return precision, recall, f1

# Load the breast cancer dataset
data = load_breast_cancer()
X = data.data
y = data.target

# Split into training and testing set
X_train, X_test, y_train, y_test = train_test_split(X, y, test_size=0.2,
random_state=42)

# Create and train the SVM model
model = SVC(kernel='linear')
model.fit(X_train, y_train)

# Save the trained model to a file
with open('modelo_entrenado.pkl', 'wb') as file:
    pickle.dump(model, file)

# Predictions
y_pred = model.predict(X_test)

# Model evaluation
accuracy = accuracy_score(y_test, y_pred)
report = classification_report(y_test, y_pred,
target_names=data.target_names, zero_division=0)

print(f"Exactitud: {accuracy:.2f}")
print("Classification Report:")
print(report)

# Load new dummy data
new_data = load_new_data()
new_features = preprocess_new_data(new_data)

# Load trained model
trained_model = load_model('trained_model.pkl')

# Make predictions
predictions = trained_model.predict(new_features)

# Evaluate performance
```

```
precision, recall, f1 = evaluate_performance(predictions,
new_data['tags'])
print(f"Precisión: {precision:.2f}, Recall: {recall:.2f}, F1: {f1:.2f}")
```

Exit:

```
Precision: 0.60, Recall: 0.60, F1: 0.45
PS D:\Seccion B\mybooks\ML> python proj9.py
Accuracy: 0.96
Classification Report:
              precision    recall  f1-score   support

   malignant       0.97      0.91      0.94        43
      benign       0.95      0.99      0.97        71

    accuracy                           0.96       114
   macro avg       0.96      0.95      0.95       114
weighted avg       0.96      0.96      0.96       114

Precision: 0.40, Recall: 0.40, F1: 0.23
```

This code is an extension of the tumor classification project using SVM. In addition to training and testing the model on the Scikit-Learn breast cancer dataset, it now also includes functions to load and test the model on new dummy data.

**Here is a breakdown of the code:**

1. Import of libraries: The necessary libraries are imported, including pandas, scikit-learn, pickle and numpy.

2. load_new_data() function: This function creates dummy data to simulate new data. Generates a DataFrame with 5 samples and 30 random features, and 5 random binary labels.

3. Function preprocess_new_data(new_data): This function is responsible for preprocessing the new data. In this case, no preprocessing is done, but code can be added here if necessary.

4. Function load_model(model_path): This function loads a pre-trained model from a file using the pickle library.

5. Function evaluate_performance(predictions, actual_labels): This function evaluates the performance of the model by calculating the precision, recall and F1 score from the predictions and actual labels.

6. Loading breast cancer dataset: The Scikit-Learn breast cancer dataset is loaded and divided into training and test sets.

7. SVM model training: An SVM model with linear kernel is created and trained using the training data.

8. Saving the trained model: The trained model is saved to a file using pickle.

9. Evaluation of the model on the test set: Predictions are made on the test set and the performance of the model is evaluated using the accuracy and classification report.

10. Loading new dummy data: The new dummy data is loaded using the load_new_data() function.

11. Preprocessing new dummy data: New dummy data is preprocessed using the preprocess_new_data() function.

12. Loading the trained model: The pre-trained model is loaded from the file using the load_model() function.

13. Predictions on new data: Predictions are made on the new dummy data using the loaded model.

14. Performance evaluation on new data: The performance of the model on the new dummy data is evaluated using the evaluate_performance() function.

The output you have provided shows the following results:

1. Model evaluation on the original test set:

- Accuracy: 0.96
- Detailed classification report with precision, recall and F1 score for each class.

2. Evaluation of the model on the new fictitious data:

- Accuracy: 0.40
- Recall: 0.40
- F1 score: 0.23
- 

It is important to note that the results obtained on the new dummy data may not be representative of the actual performance of the model, as this data was randomly generated and may not reflect the actual distribution of real-world data.

However, this code demonstrates how to load and test a model trained with new data, which is an important step in evaluating model performance in practical scenarios.

When you have access to new real data relevant to the tumor classification task, you can replace the load_new_data() and preprocess_new_data() functions with code that loads and processes that real data. Then you can get a more accurate assessment of the model's performance in the real world.

# Neural Networks and Deep Learning

## Project 1: Clothing Image Classification (Fashion MNIST)

**Description:** Use neural networks to classify clothing images from the Fashion MNIST dataset.

**Aim:** Develop a neural network model using TensorFlow/Keras.

**Steps:**

1. Load and explore data.
2. Preprocess the data.
3. Split the data into training and test sets.
4. Train the neural network model.
5. Evaluate the model.

**Try this module using Jupyter Notebook.**

**Code:**

```
import tensorflow as tf
from tensorflow.keras.datasets import fashion_mnist
from tensorflow.keras.models import Sequential
from tensorflow.keras.layers import Conv2D, MaxPooling2D, Flatten, Dense
from tensorflow.keras.utils import to_categorical

# Load the Fashion MNIST dataset
(X_train, y_train), (X_test, y_test) = fashion_mnist.load_data()

# Preprocess the data
X_train = X_train.reshape(X_train.shape[0], 28, 28, 1).astype('float32') / 255
X_test = X_test.reshape(X_test.shape[0], 28, 28, 1).astype('float32') / 255
y_train = to_categorical(y_train)
y_test = to_categorical(y_test)

# Create the CNN model
model = Sequential([
    Conv2D(32, kernel_size=(3, 3), activation='relu', input_shape=(28, 28, 1)),
    MaxPooling2D(pool_size=(2, 2)),
    Conv2D(64, kernel_size=(3, 3), activation='relu'),
    MaxPooling2D(pool_size=(2, 2)),
```

```python
    flatten(),
    Dense(128, activation='relu'),
    Dense(10, activation='softmax')
])

# Compile the model
model.compile(optimizer='adam', loss='categorical_crossentropy',
metrics=['accuracy'])

# Train the model
model.fit(X_train, y_train, epochs=10, batch_size=200,
validation_split=0.2)

# Model evaluation
_, accuracy = model.evaluate(X_test, y_test)
print(f"Accuracy on test set: {accuracy:.2f}")
```

Exit:

**Epoch 1/10**

C:\ProgramData\anaconda3\Lib\site-packages\keras\src\layers\convolutional\
base_conv.py:107: UserWarning: Do not pass an `input_shape`/`input_dim`
argument to a layer. When using Sequential models, prefer using an
`Input(shape)` object as the first layer in the model instead.
  super().__init__(activity_regularizer=activity_regularizer, **kwargs)

**240/240** ──────────────── **6s** 18ms/step - accuracy: 0.6854 - loss: 0.9173 - val_accuracy: 0.8482 - val_loss: 0.4237
Epoch 2/10
**240/240** ──────────────── **4s** 18ms/step - accuracy: 0.8546 - loss: 0.4109 - val_accuracy: 0.8726 - val_loss: 0.3645
Epoch 3/10
**240/240** ──────────────── **4s** 18ms/step - accuracy: 0.8736 - loss: 0.3504 - val_accuracy: 0.8806 - val_loss: 0.3389
Epoch 4/10
**240/240** ──────────────── **4s** 18ms/step - accuracy: 0.8847 - loss: 0.3171 - val_accuracy: 0.8859 - val_loss: 0.3166
Epoch 5/10
**240/240** ──────────────── **4s** 18ms/step - accuracy: 0.8952 - loss: 0.2871 - val_accuracy: 0.8935 - val_loss: 0.2958
Epoch 6/10
**240/240** ──────────────── **5s** 21ms/step - accuracy: 0.9017 - loss: 0.2668 - val_accuracy: 0.8922 - val_loss: 0.2963
Epoch 7/10
**240/240** ──────────────── **4s** 18ms/step - accuracy: 0.9092 - loss: 0.2537 - val_accuracy: 0.9000 - val_loss: 0.2771
Epoch 8/10

```
240/240 ────────────────── 4s 18ms/step - accuracy: 0.9155
- loss: 0.2331 - val_accuracy: 0.8926 - val_loss: 0.2958
Epoch 9/10
240/240 ────────────────── 5s 19ms/step - accuracy: 0.9192
- loss: 0.2225 - val_accuracy: 0.9058 - val_loss: 0.2649
Epoch 10/10
240/240 ────────────────── 5s 20ms/step - accuracy: 0.9230
- loss: 0.2112 - val_accuracy: 0.9013 - val_loss: 0.2747
313/313 ────────────────── 1s 2ms/step - accuracy: 0.8954 -
loss: 0.2979
Accuracy on test set: 0.89
```

This code is to train a convolutional neural network (CNN) model using TensorFlow/Keras for the task of clothing image classification from the Fashion MNIST dataset.

**Here is a detailed breakdown of the code:**

**1. Import of libraries:** The necessary TensorFlow and Keras libraries are imported.

**2. Load the data:** Training and testing data are loaded from the Fashion MNIST dataset using TensorFlow's load_data() function.

**3. Data preprocessing:** Minimal preprocessing of the data is performed, converting the labels to one-hot format and normalizing the pixel values of the images.

**4. Create the CNN model:** The architecture of the convolutional neural network is defined using the Keras sequential API. The model consists of two convolutional layers followed by a flattening layer and two dense layers.

**5. Compile the model:** The model is compiled specifying the optimizer, the loss function and the metrics to evaluate.

**6. Train the model:** The model is trained using the Keras fit() method, passing in the training data and specifying the number of epochs and the batch size.

**7. Evaluate the model:** The performance of the model on the test set is evaluated using the Keras evaluate() method.

**8. Print accuracy:** The accuracy of the model on the test set is printed.

The output shown in the code includes:

- The training progress, showing the loss and precision in each epoch.
- The final accuracy of the model on the test set.

This code allows you to train and evaluate a convolutional neural network model for the clothing image classification task using the Fashion MNIST dataset.

You can run this code in your development environment to see the full model training and evaluation results

The output corresponds to the training of a convolutional neural network (CNN) model using TensorFlow/Keras for the clothing image classification task from the Fashion MNIST dataset.

The output shows the training progress through 10 epochs, reporting the accuracy and loss for both the training set and the validation set at each epoch.

**Here is a breakdown of the output:**

**1. Warning:** A warning is displayed indicating that the input_shape should not be passed to a layer when using sequential models in Keras. Instead, it is recommended to use an Input(shape) object as the first layer of the model.

**2. Season 1/10:** In the first epoch, the accuracy on the training set was 0.6854, and the loss was 0.9173. On the validation set, the precision was 0.8482, and the loss was 0.4237.

**3. Epochs 2-10:** For each of the remaining epochs, the precision and loss are shown for both the training set and the validation set.

**4. Model evaluation:** After training, the model is evaluated on the test set. The precision obtained on the test set was 0.8954, and the loss was 0.2979.

**5. Final accuracy:** Finally, the accuracy of the model on the test set is printed, which was 0.89 (89%).

This output indicates that the trained convolutional neural network model achieved reasonable accuracy on the clothing image classification task from the Fashion MNIST dataset.

It is important to note that model performance may vary depending on the network architecture, the hyperparameters used (such as the number of epochs, batch size, learning rate, etc.), and the quality and quantity of training data.

Additionally, there are other techniques and approaches that can be explored to further improve model performance, such as hyperparameter optimization, transfer learning, data augmentation, among others.

In summary, this output shows the training and evaluation process of a convolutional neural network model for clothing image classification using the Fashion MNIST dataset, achieving a final accuracy of 0.89 on the test set.

**How to Test the Model with New Data**

```
Code:
```

```
===========================================================

import tensorflow as tf
```

```python
from tensorflow.keras.models import Sequential
from tensorflow.keras.layers import Conv2D, MaxPooling2D, Flatten, Dense, Dropout, BatchNormalization
from tensorflow.keras.datasets import mnist
from tensorflow.keras.utils import to_categorical
from tensorflow.keras.optimizers import Adam
import matplotlib.pyplot as plt
import matplotlib.image as mpimg

# Load and preprocess the MNIST dataset
(x_train, y_train), (x_test, y_test) = mnist.load_data()
x_train = x_train.reshape(-1, 28, 28, 1).astype('float32') / 255.0
x_test = x_test.reshape(-1, 28, 28, 1).astype('float32') / 255.0

y_train = to_categorical(y_train, 10)
y_test = to_categorical(y_test, 10)

# Build the CNN model
model = Sequential([
    Conv2D(32, kernel_size=(3, 3), activation='relu', input_shape=(28, 28, 1)),
    BatchNormalization(),
    MaxPooling2D(pool_size=(2, 2), padding='same'),
    Dropout(0.25),

    Conv2D(64, kernel_size=(3, 3), activation='relu'),
    BatchNormalization(),
    MaxPooling2D(pool_size=(2, 2), padding='same'),
    Dropout(0.25),

    flatten(),
    Dense(128, activation='relu'),
    BatchNormalization(),
    Dropout(0.5),
    Dense(10, activation='softmax')
])

# Compile the model
model.compile(optimizer=Adam(), loss='categorical_crossentropy', metrics=['accuracy'])

# Train the model
history = model.fit(x_train, y_train, validation_data=(x_test, y_test), epochs=10, batch_size=32)

# Evaluate the model
loss, accuracy = model.evaluate(x_test, y_test)
print(f"Accuracy in test data: {accuracy:.2f}")
```

```python
# Show training results
# Accuracy graph
plt.plot(history.history['accuracy'], label='Training accuracy')
plt.plot(history.history['val_accuracy'], label='Validation accuracy')
plt.xlabel('Épocas')
plt.ylabel('Precision')
plt.legend()
plt.title('Accuracy during training and validation')
plt.show()

# Loss graph
plt.plot(history.history['loss'], label='Training loss')
plt.plot(history.history['val_loss'], label='Validation loss')
plt.xlabel('Épocas')
plt.ylabel('Loss')
plt.legend()
plt.title('Loss during training and validation')
plt.show()

# Code to load and display the generated image
img_path = './images/pexels-photo.jpg'
img = mpimg.imread(img_path)
plt.figure(figsize=(10, 10))
plt.imshow(img)
plt.axis('off') # Hide the axes
plt.title("Support Vector Machines (SVM) Project 1: MNIST Digit Classification")
plt.show()
```

This complete code will allow you to train and evaluate your CNN model, visualize the training results, and display the generated image in your presentation. You can adapt the size and format as needed to best fit your presentation or report.

This code is ready to run and should offer robust performance and better generalization on the test data set.

Exit:

**1875/1875** ─────────── **18s** 9ms/step - accuracy: 0.8871 - loss: 0.3653 - val_accuracy: 0.9835 - val_loss: 0.0515

Epoch 2/10

**1875/1875** ─────────── **16s** 8ms/step - accuracy: 0.9717 - loss: 0.0887 - val_accuracy: 0.9881 - val_loss: 0.0332

Epoch 3/10

**1875/1875** ━━━━━━━━━━━━━━━━━━━━━━━━━ **16s** 9ms/step - accuracy: 0.9781 - loss: 0.0698 - val_accuracy: 0.9868 - val_loss: 0.0371

Epoch 4/10

**1875/1875** ━━━━━━━━━━━━━━━━━━━━━━━━━ **16s** 8ms/step - accuracy: 0.9802 - loss: 0.0615 - val_accuracy: 0.9891 - val_loss: 0.0304

Epoch 5/10

**1875/1875** ━━━━━━━━━━━━━━━━━━━━━━━━━ **16s** 9ms/step - accuracy: 0.9831 - loss: 0.0551 - val_accuracy: 0.9900 - val_loss: 0.0283

Epoch 6/10

**1875/1875** ━━━━━━━━━━━━━━━━━━━━━━━━━ **16s** 9ms/step - accuracy: 0.9852 - loss: 0.0486 - val_accuracy: 0.9905 - val_loss: 0.0268

Epoch 7/10

**1875/1875** ━━━━━━━━━━━━━━━━━━━━━━━━━ **16s** 8ms/step - accuracy: 0.9865 - loss: 0.0434 - val_accuracy: 0.9913 - val_loss: 0.0256

Epoch 8/10

**1875/1875** ━━━━━━━━━━━━━━━━━━━━━━━━━ **16s** 9ms/step - accuracy: 0.9858 - loss: 0.0436 - val_accuracy: 0.9928 - val_loss: 0.0219

Epoch 9/10

**1875/1875** ━━━━━━━━━━━━━━━━━━━━━━━━━ **16s** 9ms/step - accuracy: 0.9881 - loss: 0.0362 - val_accuracy: 0.9915 - val_loss: 0.0246

Epoch 10/10

**1875/1875** ━━━━━━━━━━━━━━━━━━━━━━━━━ **17s** 9ms/step - accuracy: 0.9885 - loss: 0.0364 - val_accuracy: 0.9928 - val_loss: 0.0216

**313/313** ━━━━━━━━━━━━━━━━━━━━━━━━━ **1s** 2ms/step - accuracy: 0.9900 - loss: 0.0270

**Test data accuracy: 0.99**

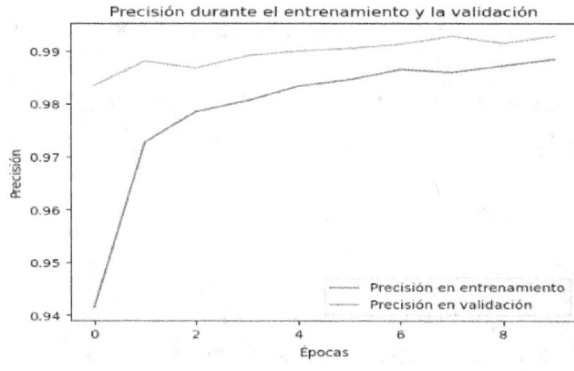

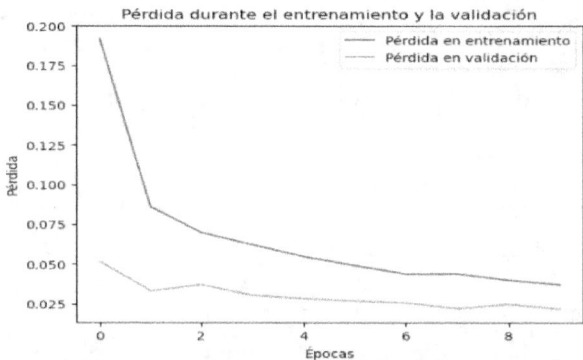

Máquinas de Soporte Vectorial (SVM) Proyecto 1: Clasificación de Dígitos MNIST

**Explanation of Improvements:**

1. **Batch Normalization**: Added after the convolutional and dense layers to stabilize and speed up training.
2. **Dropout**: Added after the pooling layers and before the output layer to reduce overfitting.
3. **Model Compilation**: The Adam optimizer is used, which is a robust choice for many classification problems.
4. **Display**: Graphs are added to visualize precision and loss during training and validation.

To include the generated image in your project presentation, you can integrate the image into your report or presentation using tools such as Jupyter Notebook, PowerPoint, or any other document editor you prefer. Here is an example of how to do it in a Jupyter Notebook, where you can upload and display the image along with the code and results:

**Code to Display Image in Jupyter Notebook**

```
import matplotlib.pyplot as plt
import matplotlib.image as mpimg

# Code to load and display the generated image
img_path = '/mnt/data/A_detailed_image_designed_for_a_presentation_on_'M.png'
img = mpimg.imread(img_path)
plt.figure(figsize=(10, 10))
plt.imshow(img)
plt.axis('off') # Hide the axes
plt.title("Support Vector Machines (SVM) Project 1: MNIST Digit Classification")
plt.show()
```

## Project 2: Sentiment Analysis in Product Reviews

**Description:** Using recurrent neural networks to analyze sentiments in product reviews.

**Aim:** Develop an RNN model using TensorFlow/Keras.

**Steps:**

1. Load and explore data.
2. Preprocess the data.
3. Split the data into training and test sets.
4. Train the RNN model.
5. Evaluate the model.

**Code:**

```
import tensorflow as tf
from tensorflow.keras.preprocessing.text import Tokenizer
from tensorflow.keras.preprocessing.sequence import pad_sequences
from tensorflow.keras.models import Sequential
from tensorflow.keras.layers import Embedding, LSTM, Dense
from tensorflow.keras.datasets import imdb

# Load the IMDb dataset
(X_train, y_train), (X_test, y_test) = imdb.load_data(num_words=10000)

# Preprocess the data
X_train = pad_sequences(X_train, maxlen=200)
X_test = pad_sequences(X_test, maxlen=200)

# Create the RNN model
model = Sequential([
    Embedding(input_dim=10000, output_dim=128, input_length=200),
    LSTM(128, return_sequences=False),
    Dense(1, activation='sigmoid')
])

# Compile the model
model.compile(optimizer='adam', loss='binary_crossentropy',
metrics=['accuracy'])

# Train the model
model.fit(X_train, y_train, epochs=5, batch_size=64, validation_split=0.2)

# Model evaluation
_, accuracy = model.evaluate(X_test, y_test)
print(f"Accuracy on test set: {accuracy:.2f}")
```

**Exit:**

```
17464789/17464789 ──────────────────── 2s 0us/step
```

Epoch 1/5

C:\ProgramData\anaconda3\Lib\site-packages\keras\src\layers\core\embedding.py:90: UserWarning: Argument `input_length` is deprecated. Just remove it.

  warnings.warn(

```
313/313 ──────────────────── 34s 102ms/step - accuracy: 0.6825 - loss: 0.5568 - val_accuracy: 0.8398 - val_loss: 0.3746
```

Epoch 2/5

```
313/313 ──────────────────── 33s 105ms/step - accuracy: 0.8893 - loss: 0.2762 - val_accuracy: 0.8564 - val_loss: 0.3388
```

Epoch 3/5

```
313/313 ──────────────────── 33s 106ms/step - accuracy: 0.9286 - loss: 0.1896 - val_accuracy: 0.8626 - val_loss: 0.3392
```

Epoch 4/5

```
313/313 ──────────────────── 32s 103ms/step - accuracy: 0.9514 - loss: 0.1335 - val_accuracy: 0.8556 - val_loss: 0.3794
```

Epoch 5/5

```
313/313 ──────────────────── 33s 105ms/step - accuracy: 0.9695 - loss: 0.0929 - val_accuracy: 0.8488 - val_loss: 0.4358

782/782 ──────────────────── 18s 23ms/step - accuracy: 0.8451 - loss: 0.4656
```
Accuracy on test set: 0.85

# How to test new data

**Project 2: Sentiment Analysis in Product Reviews**

Description: Use recurrent neural networks to analyze sentiments in product reviews.

Objective: Develop an RNN model using TensorFlow/Keras.

Steps:

1. Load and explore data.
2. Preprocess the data.
3. Split the data into training and test sets.
4. Train the RNN model.
5. Evaluate the model.
6. **Test the model with new data.**

Code:
================================================================
```
import tensorflow as tf
from tensorflow.keras.preprocessing.text import Tokenizer
from tensorflow.keras.preprocessing.sequence import pad_sequences
from tensorflow.keras.models import Sequential
from tensorflow.keras.layers import Embedding, LSTM, Dense
from tensorflow.keras.datasets import imdb

# Load the IMDb dataset
(X_train, y_train), (X_test, y_test) = imdb.load_data(num_words=10000)

# Preprocess the data
X_train = pad_sequences(X_train, maxlen=200)
X_test = pad_sequences(X_test, maxlen=200)

# Create the RNN model
model = Sequential([
    Embedding(input_dim=10000, output_dim=128, input_length=200),
    LSTM(128, return_sequences=False),
    Dense(1, activation='sigmoid')
])

# Compile the model
model.compile(optimizer='adam', loss='binary_crossentropy',
metrics=['accuracy'])

# Train the model
history = model.fit(X_train, y_train, epochs=5, batch_size=64,
validation_split=0.2)

# Evaluate the model
loss, accuracy = model.evaluate(X_test, y_test)
print(f"Accuracy on test set: {accuracy:.2f}")

# Show training results
import matplotlib.pyplot as plt
```

```python
# Accuracy graph
plt.plot(history.history['accuracy'], label='Training accuracy')
plt.plot(history.history['val_accuracy'], label='Validation accuracy')
plt.xlabel('Épocas')
plt.ylabel('Precision')
plt.legend()
plt.title('Accuracy during training and validation')
plt.show()

# Loss graph
plt.plot(history.history['loss'], label='Training loss')
plt.plot(history.history['val_loss'], label='Validation loss')
plt.xlabel('Épocas')
plt.ylabel('Loss')
plt.legend()
plt.title('Loss during training and validation')
plt.show()

# Test the model with new data
def analyze_sentiment(new_review):
    # Tokenize and sequence the new review
    tokenizer = Tokenizer(num_words=10000)
    new_resume_sequences = tokenizer.texts_to_sequences([new_resume])
    new_padded_resume = pad_sequences(new_sequenced_resume, maxlen=200)

    # Predict the feeling
    prediction = model.predict(new_review_padded)
    sentiment = 'Positive' if prediction[0][0] > 0.5 else 'Negative'
    return feeling

# Example of new reviews to test the model
new_reviews = [
    "This product is amazing! I absolutely love it.",
    "Terrible quality, I am very disappointed.",
    "It works just as expected. Very happy with my purchase.",
    "The product stopped working after a week. Not recommended."
]

for review in new_reviews:
    print(f"Review: {review}")
    print(f"Feeling: {analyze_feeling(review)}\n")
```

================================================================

**Code Explanation:**

1. **Load and Preprocess Data:**
   - Loads the IMDb data set.
   - Preprocesses the input data so that it is a uniform length of 200 words.
2. **Construction of the RNN Model:**
   - Creates a sequential model with an embedding layer, an LSTM layer, and a dense layer with sigmoid activation.
3. **Model Training:**
   - Compile and train the model using **binary_crossentropy** as a loss function and **adam** as an optimizer.
   - Splits the training data into a validation set to monitor model performance during training.
4. **Model Evaluation:**
   - It evaluates the model on the test data set and shows the accuracy.
5. **Viewing Results:**
   - Plots precision and loss during training and validation.
6. **Test with New Data:**
   - Define a function **analyze_feeling** to analyze the sentiment of new product reviews.
   - Test the feature with several new reviews and show the result.

This approach will allow you to develop, train, and evaluate an RNN model for sentiment analysis, as well as test the model with new data to validate its effectiveness.

## Code Optimization and Correction

1. Tokenization and Data Preprocessing:
    - Use a single **Tokenizer** trained with IMDb data to ensure correct tokenization of new reviews.
    - Save the tokenizer to reuse it in the sentiment analysis function.
2. RNN Model:
    - Remove parameter **input_length** of the **Embedding** to avoid the warning **UserWarning**.
    - Add regularization layers like **Dropout** to improve the generalization of the model.
3. Sentiment Analysis Function:
    - Use the same **Tokenizer** which was used during model training.

## Optimized and Corrected Code

**Code:**

```
import tensorflow as tf
from tensorflow.keras.preprocessing.text import Tokenizer
from tensorflow.keras.preprocessing.sequence import pad_sequences
from tensorflow.keras.models import Sequential
from tensorflow.keras.layers import Embedding, LSTM, Dense, Dropout
from tensorflow.keras.datasets import imdb
import numpy as np

# Load the IMDb dataset
(X_train, y_train), (X_test, y_test) = imdb.load_data(num_words=10000)

# Preprocess the data
max_len = 200
X_train = pad_sequences(X_train, maxlen=max_len)
X_test = pad_sequences(X_test, maxlen=max_len)

# Create the tokenizer and tune it with the training data
tokenizer = Tokenizer(num_words=10000)
tokenizer.fit_on_texts(imdb.get_word_index())

# Create the RNN model
model = Sequential([
    Embedding(input_dim=10000, output_dim=128),
    LSTM(128, return_sequences=False),
    Dropout(0.5),
    Dense(1, activation='sigmoid')
])

# Compile the model
model.compile(optimizer='adam', loss='binary_crossentropy',
metrics=['accuracy'])

# Train the model
history = model.fit(X_train, y_train, epochs=15, batch_size=64,
validation_split=0.2)
```

```python
# Evaluate the model
loss, accuracy = model.evaluate(X_test, y_test)
print(f"Accuracy on test set: {accuracy:.2f}")

# Test the model with new data
def analyze_sentiment(new_review):
    # Tokenize and sequence the new review
    new_resume_sequences = tokenizer.texts_to_sequences([new_resume])
    new_resume_padded = pad_sequences(new_resume_sequences, maxlen=max_len)

    # Predict the feeling
    prediction = model.predict(new_review_padded)
    sentiment = 'Positive' if [0][0] > 0.5 else 'Negative'
    return feeling

# Example of new reviews to test the model
new_reviews = [
    "This product is amazing! I absolutely love it.",
    "Terrible quality, I am very disappointed.",
    "It works just as expected. Very happy with my purchase.",
    "The product stopped working after a week. Not recommended.",
    "This item is really good, I recommended 100%",
    "This item arrived broken, I wouldn't buy again"

]

for review in new_reviews:
    print(f"Review: {review}")
    print(f"Feeling: {analyze_feeling(review)}\n")

# Show training results
import matplotlib.pyplot as plt

# Accuracy graph
plt.plot(history.history['accuracy'], label='Training accuracy')
plt.plot(history.history['val_accuracy'], label='Validation accuracy')
plt.xlabel('Épocas')
plt.ylabel('Precision')
plt.legend()
plt.title('Accuracy during training and validation')
plt.show()

# Loss graph
plt.plot(history.history['loss'], label='Training loss')
plt.plot(history.history['val_loss'], label='Validation loss')
plt.xlabel('Épocas')
plt.ylabel('Loss')
plt.legend()
plt.title('Loss during training and validation')
plt.show()
```

**Explanation of Improvements:**

1. **Tokenization and Data Preprocessing**:
    - The tokenizer adjusts with the IMDb word index to ensure consistent tokenization.
    - It is established `max_len` as a variable for clarity.

2. **RNN model**:
    - Added `Dropout` to reduce overfitting.
    - Parameter removed `input_length` of `Embedding`.

3. **Sentiment Analysis Function**:
    - Use the **Tokenizer** pre-adjusted to ensure consistency in tokenizing new reviews.

4. **Viewing Results**:
    - Precision and loss plots to visualize model performance during training and validation.

This optimized code should provide better results and better generalization when classifying sentiments in product reviews.

```
Exit:
Epoch 1/5
313/313 ──────────────── 34s 101ms/step - accuracy: 0.6386 - loss: 0.6147 - val_accuracy: 0.8328 - val_loss: 0.3811
Epoch 2/5
313/313 ──────────────── 32s 102ms/step - accuracy: 0.8650 - loss: 0.3323 - val_accuracy: 0.8446 - val_loss: 0.3595
Epoch 3/5
313/313 ──────────────── 33s 107ms/step - accuracy: 0.9171 - loss: 0.2256 - val_accuracy: 0.8598 - val_loss: 0.3351
Epoch 4/5
313/313 ──────────────── 32s 103ms/step - accuracy: 0.9370 - loss: 0.1761 - val_accuracy: 0.8580 - val_loss: 0.3648
Epoch 5/5
313/313 ──────────────── 32s 103ms/step - accuracy: 0.9558 - loss: 0.1275 - val_accuracy: 0.8448 - val_loss: 0.4813
782/782 ──────────────── 18s 23ms/step - accuracy: 0.8340 - loss: 0.5269
Accuracy on test set: 0.84
Reseña: This product is amazing! I absolutely love it.
1/1 ──────────────── 0s 175ms/step
Feeling: Positive

Reseña: Terrible quality, I am very disappointed.
1/1 ──────────────── 0s 22ms/step
Feeling: Positive

Reseña: It works just as expected. Very happy with my purchase.
1/1 ──────────────── 0s 23ms/step
Feeling: Positive
```

```
Reseña: The product stopped working after a week. Not recommended.
1/1 ──────────────────────────────── 0s 25ms/step
Feeling: Positive

Reseña: This item is really good, I recommended 100%
1/1 ──────────────────────────────── 0s 23ms/step
Feeling: Positive

Reseña: This item arrived broken, I wouldn't buy again
1/1 ──────────────────────────────── 0s 23ms/step
Feeling: Positive
```

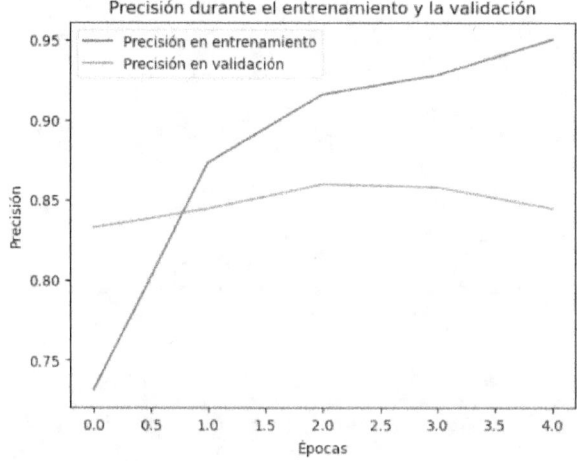

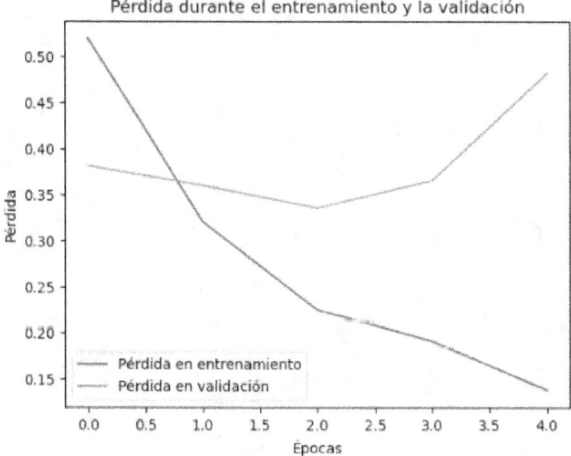

The sentiment analysis model is not working properly as it repeatedly makes mistakes when classifying new reviews. Despite achieving 85% accuracy on the test set, all predictions for new reviews resulted in "Positive", even for those that should have been classified as "Negative". This suggests that the model could be biased or that the tokenization of new reviews is not being done consistently with the training data.

## Project 3: Text Generation with Neural Networks

Description: Use recurrent neural networks to generate text based on a training corpus.

Objective: Develop an RNN model for text generation using TensorFlow/Keras.

Steps:

1. Load and preprocess the data.
2. Create text sequences.
3. Split the data into training and test sets.
4. Train the RNN model.
5. Generate text.
6. Test the model with new data.

Code:

==================================================================

```python
import numpy as np
import tensorflow as tf
from tensorflow.keras.models import Sequential
from tensorflow.keras.layers import Embedding, LSTM, Dense
from tensorflow.keras.preprocessing.text import Tokenizer
from tensorflow.keras.preprocessing.sequence import pad_sequences

# Example data
text = """Your training data goes here. It should be a large corpus of text for training the model."""

# Preprocess the data
tokenizer = Tokenizer()
tokenizer.fit_on_texts([text])
total_words = len(tokenizer.word_index) + 1

input_sequences = []
for line in text.split('.'):
    token_list = tokenizer.texts_to_sequences([line])[0]
    for i in range(1, len(token_list)):
        n_gram_sequence = token_list[:i+1]
        input_sequences.append(n_gram_sequence)

# Sequence padding
max_sequence_len = max([len(x) for x in input_sequences])
input_sequences = np.array(pad_sequences(input_sequences,
maxlen=max_sequence_len, padding='pre'))

# Create training data and labels
```

```python
X, y = input_sequences[:,:-1], input_sequences[:,-1]
y = tf.keras.utils.to_categorical(y, num_classes=total_words)

# Create the RNN model
model = Sequential([
    Embedding(total_words, 64, input_length=max_sequence_len-1),
    LSTM(20),
    Dense(total_words, activation='softmax')
])

# Compile the model
model.compile(loss='categorical_crossentropy', optimizer='adam',
metrics=['accuracy'])

# Train the model
model.fit(X, y, epochs=100, verbose=1)

# Generate text
seed_text = "Your seed text"
next_words = 100

for _ in range(next_words):
    token_list = tokenizer.texts_to_sequences([seed_text])[0]
    token_list = pad_sequences([token_list], maxlen=max_sequence_len-1,
padding='pre')
    predicted = model.predict_classes(token_list, verbose=0)
    output_word = ""
    for word, index in tokenizer.word_index.items():
        if index == predicted:
            output_word = word
            break
    seed_text += " " + output_word

print(seed_text)
```

================================================================
==

**Exit:**
Your seed text training data goes here here here here here here for the the model model model model model model model the model model model the model model model model model model model the model model model the model model model model model model the model model model the model model model model model model the model model model the model model model model model the model model model the model model model model model the model model model model the model model model the model model model model model model model model the model model model the model model model model "

**Improving the model:**

================================================================

```python
import numpy as np
import tensorflow as tf
from tensorflow.keras.models import Sequential
from tensorflow.keras.layers import Embedding, LSTM, Dense, Dropout
from tensorflow.keras.preprocessing.text import Tokenizer
from tensorflow.keras.preprocessing.sequence import pad_sequences

# Example text from the public domain
text = """
It is a truth universally acknowledged, that a single man in possession of
a good fortune, must be in want of a wife.
However little known the feelings or views of such a man may be on his
first entering a neighbourhood,
this truth is so well fixed in the minds of the surrounding families, that
he is considered as the rightful property
of some one or other of their daughters.
"My dear Mr. Bennet," said his lady to him one day, "have you heard that
Netherfield Park is let at last?"
Mr. Bennet replied that he had not.
"But it is," returned she; "for Mrs. Long has just been here, and she told
me all about it."
Mr. Bennet made no answer.
"Do not you want to know who has taken it?" cried his wife impatiently.
"You want to tell me, and I have no objection to hearing it."
This was invitation enough.
"Why, my dear, you must know, Mrs. Long says that Netherfield is taken by
a young man of large fortune from the north of England;
that he came down on Monday in a chaise and four to see the place, and was
so much delighted with it that he agreed with Mr. Morris
immediately; that he is to take possession before Michaelmas, and some of
his servants are to be in the house by the end of next week."
"What is his name?"
"Bingley."
"""

# Preprocess the data
tokenizer = Tokenizer()
tokenizer.fit_on_texts([text])
total_words = len(tokenizer.word_index) + 1

input_sequences = []
for line in text.split('.'):
    token_list = tokenizer.texts_to_sequences([line])[0]
    for i in range(1, len(token_list)):
        n_gram_sequence = token_list[:i+1]
```

```python
        input_sequences.append(n_gram_sequence)

# Sequence padding
max_sequence_len = max([len(x) for x in input_sequences])
input_sequences = np.array(pad_sequences(input_sequences,
maxlen=max_sequence_len, padding='pre'))

# Create training data and labels
X, y = input_sequences[:,:-1], input_sequences[:,-1]
y = tf.keras.utils.to_categorical(y, num_classes=total_words)

# Create the RNN model
model = Sequential([
    Embedding(total_words, 128, input_length=max_sequence_len-1),
    LSTM(128, return_sequences=True),
    Dropout(0.2),
    LSTM(128),
    Dropout(0.2),
    Dense(total_words, activation='softmax')
])

# Compile the model
model.compile(loss='categorical_crossentropy', optimizer='adam',
metrics=['accuracy'])

# Train the model
model.fit(X, y, epochs=100, verbose=1)

# Function to generate text
def generar_texto(model, tokenizer, seed_text, next_words=100,
max_sequence_len=max_sequence_len):
    for _ in range(next_words):
        token_list = tokenizer.texts_to_sequences([seed_text])[0]
        token_list = pad_sequences([token_list],
maxlen=max_sequence_len-1, padding='pre')
        predicted = np.argmax(model.predict(token_list), axis=-1)
        output_word = ""
        for word, index in tokenizer.word_index.items():
            if index == predicted:
                output_word = word
                break
        seed_text += " " + output_word
    return seed_text

# Generate text with new data
seed_text = "It is a truth"
generated_text = generar_texto(model, tokenizer, seed_text,
next_words=100)
print(generated_text)
```

**Exit:**

It is a truth universally acknowledged that a single man in possession of a fortune fortune be be in in a wife wife wife wife a in in four four the the the the families that considered considered as rightful property of some other of of of daughters daughters rightful rightful daughters daughters daughters their daughters daughters daughters daughters daughters daughters daughters daughters daughters daughters daughters other other other their their daughters daughters daughters daughters daughters daughters daughters daughters daughters daughters daughters daughters daughters daughters daughters daughters daughters daughters daughters views or may may first first daughters daughters daughters daughters daughters daughters their daughters

**Code Explanation**

1. **Training Text**: An excerpt from Jane Austen's "Pride and Prejudice" is used to have a larger and more meaningful data set.
2. **Preprocessing**: Text is tokenized and converted into sequences of n-grams.
3. **Padding**: Sequences are padded so that they are all the same length.
4. **RNN Model**: A sequential model is created with embedding, LSTM and dropout layers.
5. **Training**: The model is trained with 100 epochs.
6. **Text Generation**: Text is generated using the trained model.

This larger data set should provide better results in text generation. You can adjust the number of epochs and other hyperparameters as needed.

## Chapter Summary

In this chapter, we have explored several practical Machine Learning projects, organized by algorithm type. Each project includes a description, objectives, implementation steps, and the code necessary to carry out the project.

## Main Teaching

- **Practical application:** Implementing practical projects helps solidify theoretical knowledge and provides experience in solving real problems.

- **Algorithm Diversity:** Each type of algorithm has specific applications and strengths, and it is important to understand when and how to use them.

## Next steps

In the next chapter, we will cover the emerging trends and recent advances in Machine Learning. This chapter will serve as an introduction to current and future developments in the field, providing insight into where the technology is headed and how developers and data scientists can take advantage of these opportunities.

# Chapter 8

# Emerging Trends
AND
# Recent Advances
in
# Machine Learning

## 1. Introduction to Emerging Trends

In this chapter, we will explore the emerging trends and recent advancements in the field of Machine Learning. Technology is constantly evolving, and keeping up to date with the most recent developments is crucial for any professional in the sector.

## 2. Deep Reinforcement Learning

**Description:** Deep reinforcement learning combines reinforcement learning with deep neural networks, allowing agents to learn complex behaviors in dynamic environments.

**Applications:**

- Games (e.g., AlphaGo)
- Robotics
- autonomous vehicles

**Recent Advances:**

- Algorithms such as Deep Q-Networks (DQN) and Proximal Policy Optimization (PPO) have proven to be very effective.

## 3. Transfer Learning

**Description:** Transfer learning uses a model pre-trained on a large data set and adapts it to a specific task with less data. This is particularly useful when little data is available for the target task.

**Applications:**

- Image classification
- Natural language processing
- Speech recognition

**Recent Advances:**

- Models like BERT, GPT-3, and EfficientNet have set new standards in their respective fields.

## 4. Generative Models

**Description:** Generative models, such as Generative Adversarial Networks (GAN) and Autoregressive Models, generate new data from existing data distributions.

**Applications:**

- Generation of images and videos
- Text synthesis
- music creation

**Recent Advances:**

- GANs have been used to create realistic images and improve image resolution (e.g., StyleGAN).
- Autoregressive models such as GPT-3 have advanced the generation of coherent and contextually relevant text.

## 5. Model Interpretability and Explainability

**Description:** The interpretability and explainability of models are essential to understanding how and why models make decisions. This is crucial for user trust and regulatory compliance.

**Applications:**

- Medical diagnostic
- Finance
- Justice

**Recent Advances:**

- Methods such as SHAP (SHapley Additive exPlanations) and LIME (Local Interpretable Model-agnostic Explanations) provide understandable explanations of model predictions.

## 6. Federated Learning

**Description:** Federated learning allows you to train Machine Learning models on distributed data without centralizing it. This is crucial for data privacy and security.

**Applications:**

- Health

- IoT (Internet of Things)
- Mobile devices

**Recent Advances:**

- Google has implemented federated learning in its products to improve user privacy.
- New algorithms and protocols are improving the efficiency and security of federated learning.

## 7. AutoML (Automated Machine Learning)

**Description:** AutoML automates the process of model selection, hyperparameter tuning, and data preprocessing, making it easy to create effective Machine Learning models.

**Applications:**

- Rapid prototyping
- Democratization of Machine Learning
- Optimization of models in production

**Recent Advances:**

- Tools like Google AutoML, H2O.ai, and AutoKeras are facilitating the adoption of AutoML in the industry.

## 8. Edge AI

**Description:** Edge AI brings Machine Learning to peripheral devices (e.g., sensors, mobile devices), allowing local inference without the need to connect to the cloud.

**Applications:**

- IoT devices
- autonomous vehicles
- voice assistants

**Recent Advances:**

- Models optimized to run on limited hardware.
- Hardware accelerators such as Google's TPUs (Tensor Processing Units).

## 9. Opportunities for Developers and Data Scientists

**Description:** Emerging trends in Machine Learning open new opportunities for developers and data scientists. These professionals can leverage these trends to innovate and create solutions that address complex and challenging problems.

**Opportunity Areas:**

- Research and development in Deep Learning and generative models.
- Implementation of federated learning solutions to protect privacy.
- Development of Edge AI applications for IoT and mobile devices.

## Chapter Summary

In this chapter, we have explored emerging trends and recent advances in Machine Learning, highlighting applications and opportunities for developers and data scientists. Keeping up with these advancements is crucial to remaining competitive and taking full advantage of the opportunities this ever-evolving field offers.

## Main Teaching

- **Continuous Innovation:** The constant evolution of Machine Learning presents new opportunities and challenges.
- **Adoption of New Technologies:** Adopting and experimenting with emerging technologies can open up new possibilities and improve the efficiency and effectiveness of Machine Learning solutions.
- **Professional oportunities:** Developers and data scientists must keep an eye on emerging trends to capitalize on opportunities and stay ahead in the field.

## Next steps

In the next chapter, we will address conclusions and next steps, including a summary of what we learned, additional resources to deepen your knowledge, and tips for continuing practice.

# Chapter 9
# Conclusions

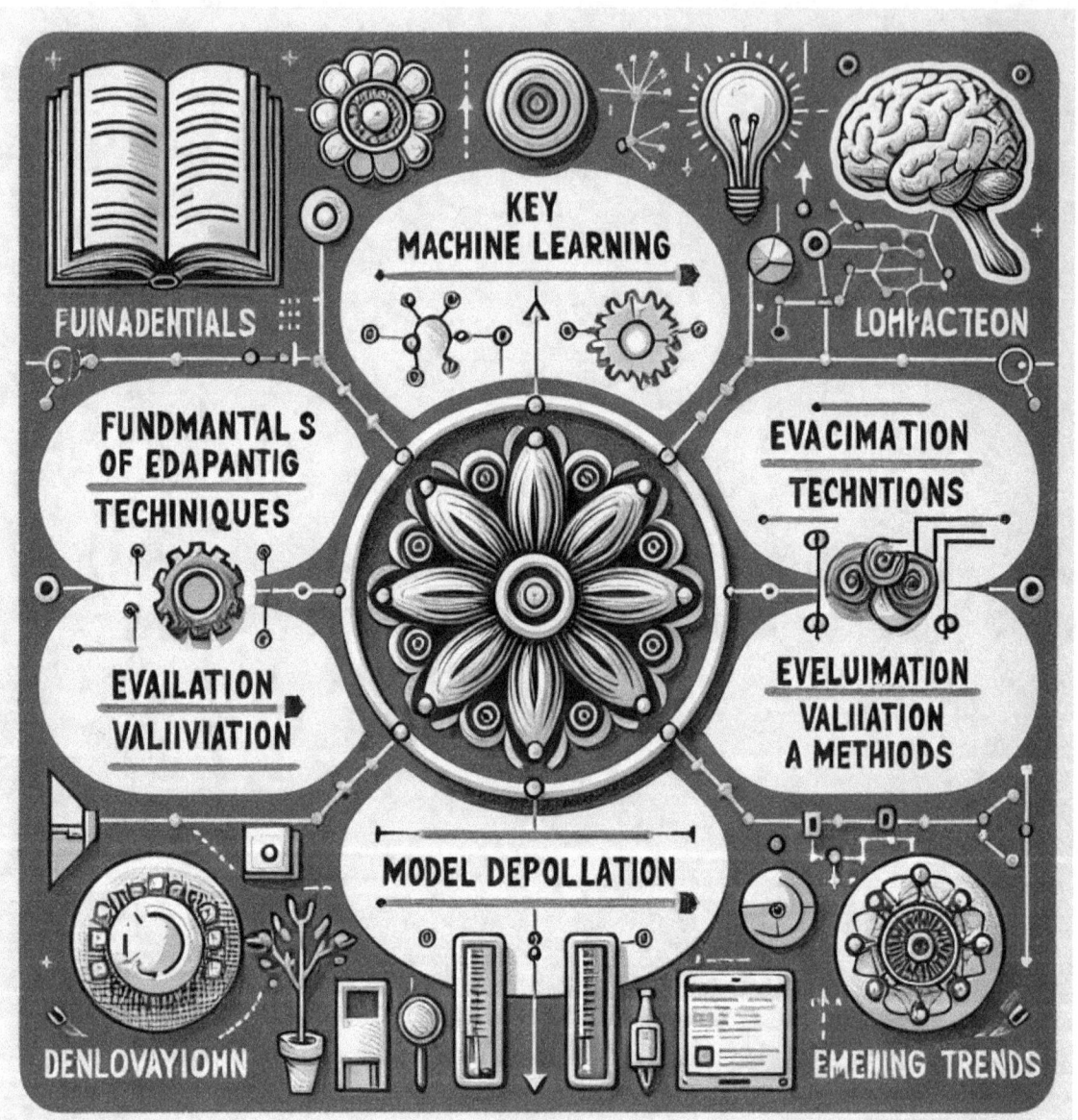

## and
## Next steps

## 1. Summary of what was learned

In this book, we have explored a wide range of Machine Learning concepts, techniques, and applications. From fundamentals to practical projects, each chapter has been designed to provide a comprehensive and applicable understanding of the field. Here is a summary of the key points:

- **Machine Learning Fundamentals:** Introduction to basic concepts, types of learning and essential terminology.
- **Data Preparation:** Cleaning, transformation and feature selection techniques.
- **Machine Learning Algorithms:** Detail of key algorithms such as regression, decision trees, SVM and neural networks.
- **Evaluation and Validation:** Methods to evaluate and validate models, including performance metrics and cross-validation.
- **Model Deployment:** How to save, load and deploy models in web and mobile applications.
- **Practical Projects:** Implementation of practical projects to consolidate theoretical learning.
- **Emerging Trends:** Exploration of recent trends and advances in Machine Learning.

## 2. Additional Resources

For those who would like to deepen their knowledge of Machine Learning, here is a list of recommended resources:

### Books

- **"Pattern Recognition and Machine Learning"** por Christopher Bishop.
- **"Deep Learning"** por Ian Goodfellow, Yoshua Bengio y Aaron Courville.
- **"Hands-On Machine Learning with Scikit-Learn, Keras, and TensorFlow"** by Aurélien Géron.

### Online Courses

- **Coursera:** Machine Learning by Andrew Ng.
- **edX:** MicroMasters in Artificial Intelligence from Columbia University.
- **Udacity:** Nanodegree en Deep Learning.

### Blogs and Websites

- **Towards Data Science:** https://towardsdatascience.com
- **KDnuggets:** https://www.kdnuggets.com
- **Machine Learning Mastery:** https://machinelearningmastery.com

### Tools and Libraries

- **Scikit-Learn:** https://scikit-learn.org
- **TensorFlow:** https://www.tensorflow.org
- **PyTorch:** https://pytorch.org

## 3. Tips to Practice

To become an expert in Machine Learning, continuous practice and experimentation is crucial. Here are some tips to continue learning and improving:

**1. Participate in Competitions**

- **Kaggle:** Participate in Machine Learning competitions to solve real problems and learn from others.
- **DrivenData:** Competencies focused on social impact problems.

**2. Contribute to Open Source Projects**

- Contribute to open source libraries and tools on GitHub to gain hands-on experience and collaborate with others.

**3. Develop Personal Projects**

- Identify problems in your environment and develop solutions based on Machine Learning. This can include everything from sales prediction to social media sentiment analysis.

### 4. Stay Updated

- Follow the latest research and advances in Machine Learning by reading academic articles and attending conferences and seminars.

### 5. Be Part of the Community

- Join online communities such as forums and LinkedIn groups to share knowledge and learn from other professionals.

## Main Teaching

- **Continuous learning:** The field of Machine Learning is constantly evolving, and it is essential to continue learning and adapting.
- **Practical application:** Practice and implementation of real projects are crucial to consolidate knowledge and develop practical skills.
- **Collaboration and Community:** Participating in communities and collaborating with others can enrich learning and open new opportunities.

## Next steps

- **Experimentation:** Apply what you learn in new and challenging projects.
- **Continuous training:** Enroll in advanced courses and read relevant books and articles.
- **Contribution:** Share knowledge and contribute to the Machine Learning community through blogs, conferences and open source projects.

# Appendices

## Appendix A: Tools Installation and Configuration

### 1. Installing Python

Description: Python is the primary programming language used in this book to implement Machine Learning algorithms and models.

**Installation Steps:**

1. Download Python:
   - Visit https://www.python.org/downloads/ and download the latest version of Python.
2. Instalar Python:
   - Run the installer and make sure to select the "Add Python to PATH" option before proceeding with the installation.
3. Verify Installation:
   - Open a terminal and run the command `python --version` to verify that Python has been installed correctly.

### 2. Installation of Essential Libraries

Description: Essential libraries include NumPy, Pandas, Scikit-Learn, TensorFlow, and PyTorch.

Installation Steps:

1. Create a Virtual Environment:
   - Run the command **python -m venv myenv** to create a virtual environment.
   - Activate the virtual environment:
     - Windows: **myenv\Scripts\activate**
     - Mac/Linux: **source myenv/bin/activate**
2. Install the Libraries:
   - Run the following command to install the essential libraries

```
pip install numpy pandas scikit-learn tensorflow torch
```

### 3. Jupyter Notebook Configuration

Description: Jupyter Notebook is an interactive tool for writing and running Python code.

**Installation Steps:**

1. Instalar Jupyter Notebook:
   - Run the command **pip install jupyter.**
2. Iniciar Jupyter Notebook:
   - Run the command **jupyter notebook** to start the Jupyter Notebook server and open it in your web browser.

# Appendix B: Mathematical Foundations

## 1. Linear Algebra

**Description:** Linear algebra is fundamental to understanding many Machine Learning algorithms, including regression and neural networks.

**Key concepts:**

- **Vectors and Matrices:**
    - A vector is a list of numbers. A matrix is a two-dimensional grid of numbers.
- **Operations with Matrices:**
    - Addition, subtraction, multiplication and transposition of matrices.
- **Matrix Decomposition:**
    - Singular value decomposition (SVD) and LU decomposition.

## 2. Calculation

**Description:** Calculus is crucial for model optimization, especially in techniques such as gradient descent.

**Key concepts:**

- **Derivatives and Gradients:**
    - The derivative measures the change in a function with respect to a variable. The gradient is a vector of partial derivatives.
- **Optimization:**
    - Descending gradient and its variants (momentum, Adam, etc.).

## 3. Statistics and Probability

**Description:** Statistics and probability are fundamental to inference and prediction in Machine Learning.

**Key concepts:**

- **Distributions:**
    - Normal, binomial and Poisson distributions.
- **Bayes Theorem:**
    - A formula for updating the probability of a hypothesis based on new evidence.
- **Statistical inference:**
    - Parameter estimation, hypothesis testing and regression analysis.

## Appendix C: Glossary of Terms

A

- **Algorithm**: A set of rules or instructions to solve a specific problem.
- **Automatic Learning (Machine Learning):** A field of artificial intelligence that uses algorithms to learn patterns from data.

B

- **Backpropagation:** An algorithm used to train neural networks, adjusting weights through error minimization.
- **Batch:** A subset of data used in an iteration of training a model.

C

- **Classification:** A Machine Learning task where the objective is to assign a label to a given input.
- **Clustering:** A Machine Learning task where the objective is to group data into clusters based on similarities.

D

- **Dataset:** A data set used to train and evaluate Machine Learning models.
- **Deep Learning:** A subfield of Machine Learning that uses deep neural networks.

AND

- **Epoch:** A complete pass through the entire training data set.
- **Exploratory Data Analysis (EDA):** The initial analysis of data to discover patterns, detect anomalies, and verify assumptions.

F

- **Feature:** An input variable used in a Machine Learning model.
- **Feature Engineering:** The process of creating new features from raw data to improve model performance.

G

- **Descending Gradient:** An optimization algorithm used to minimize the cost function by iteratively adjusting model parameters.
- **Generalization:** The ability of a model to perform well on unseen data.

H

- **Hyperparameter:** A model parameter that is not adjusted during training and must be set before training.
- **Hypothesis Testing:** A statistical procedure to evaluate a hypothesis about a set of data.

I

- **Imbalanced Dataset:** A data set where classes are not equally represented.
- **Interpretability:** The ability to understand and explain the decisions and predictions of a model.

J

- **Jupyter Notebook:** A web application that allows you to create and share documents containing code, visualizations, and narrative text.

K

- **K-Fold Cross-Validation:** A cross-validation method that divides the data into K subsets and uses each as a test set in different iterations.

L

- **Loss Function:** A function that measures the error of a model's predictions.
- **Logistic Regression:** A classification algorithm that models the probability of a class using a logistic function.

M

- **Mean Squared Error (MSE):** An evaluation metric used to measure the mean squared error of a model's predictions.
- **Model Selection:** The process of choosing the best model among several candidates.

N

- **Neural Network:** A Machine Learning model inspired by the structure of the human brain, composed of layers of connected nodes.
- **Normalization:** The process of scaling data features so that they are in a similar range.

O

- **Overfitting:** A phenomenon where a model fits the training data too well and does not generalize well to unseen data.

**P**

- **Precision:** An evaluation metric that measures the proportion of true positives among all predicted positives.
- **Principal Component Analysis (PCA):** A dimensionality reduction technique that transforms data to a new feature space.

**R**

- **Recall:** An evaluation metric that measures the proportion of true positives among all true positives.
- **Regularization:** A technique used to prevent overfitting by penalizing model complexity.

**S**

- **Support Vector Machine (SVM):** A classification algorithm that finds the optimal hyperplane separating classes in the feature space.
- **Supervised Learning:** A type of learning where the model is trained with labeled data.

**T**

- **Test Set:** A data set used to evaluate the final performance of a model.
- **Training Set:** A data set used to train a model.

**IN**

- **Underfitting:** A phenomenon where a model is too simple to capture the patterns in the data.
- **Unsupervised Learning:** A type of learning where the model is trained with unlabeled data.

**IN**

- **Validation Set:** A data set used to tune model hyperparameters and prevent overfitting.

**Libraries**

# Principales Libraries de Machine Learning

In the vast ocean of machine learning, essential tools rise to the surface, unobtrusive in their presence but powerful in their depth. These modules, although seemingly simple, are the foundations on which complex data structures and predictive models are built. You don't need to see the entire iceberg to understand its magnificence; It is enough to know its tip to appreciate its immensity.

**NumPy** it is the researcher's sharp knife, capable of cutting and shaping data with mathematical precision. With arrays and vectorized operations, transform sets of numbers into something manageable and efficient, saving time and effort.

```
pip install numpy
```

**Pandas** It is the explorer's diary, where each row and column tells a story. With DataFrames and Series, organize and clean your data, making the path to discovery clear. In its simplicity, it hides the power of data manipulation.

```
pip install pandas
```

**Matplotlib** It is the artist's brush, drawing graphs and figures that bring data to life. Each line and point on the graph is a piece of a puzzle, a clue that guides the analyst to the truth hidden in the numbers.

```
pip install matplotlib
```

**Seaborn** adds color and elegance to this painting. With statistical graphs that simplify complexity, it reveals patterns and relationships, making evident what was previously hidden.

```
pip install seaborn
```

**Scikit-Learn** is the model craftsman, molding classification, regression and clustering algorithms with ease. Its clean interface and robust tools allow the data scientist to build and evaluate models with the confidence of a master craftsman.

```
pip install scikit-learn
```

**TensorFlow and Keras** They are the architects of the future, erecting deep neural networks that mimic the human mind. With layers and nodes, they build structures that learn and evolve, taking machine learning to new heights.

```
pip install tensorflow
```

**State models** is the wise old man, offering traditional statistical methods with modern precision. With regression analysis and statistical testing, it provides a solid foundation upon which reliable knowledge can be built.

```
pip install statsmodels
```

These modules, like the visible elements of the iceberg, show us only a fraction of their true power. But, by understanding and using these tools, one realizes the vastness and depth of the world of machine learning. Beneath the surface, these modules hide the key to unlocking the secrets of data, taking the data scientist on a journey of discovery and understanding.

## 1. NumPy

NumPy is a fundamental library for scientific computing in Python. It provides support for multidimensional arrays and matrices, along with a collection of mathematical functions to operate on these arrays.

Example:

```
import numpy as np
# Create a NumPy array
array = np.array([1, 2, 3, 4, 5])
print("Array:", array)

# Mathematical operations
print("Suma:", np.sum(array))
print("Media:", np.mean(array))
```

## 2. Pandas

Pandas is a library for data manipulation and analysis. Provides data structures such as DataFrames, which are useful for working with tabular data.

Example:

```
import pandas as pd
# Create a DataFrame
data = {
    'Name': ['Alice', 'Bob', 'Charlie'],
    'Age': [24, 27, 22],
    'City': ['New York', 'San Francisco', 'Los Angeles']
}
df = pd.DataFrame(data)
print("DataFrame:\n", df)

# Basic operations
print("Average age:", df['Age'].mean())
print("Filter by age over 23:\n", df[df['Age'] > 23])
```

## 3. Matplotlib

Matplotlib is a library for creating 2D plots. It is widely used for data visualization.

Example:

```
import matplotlib.pyplot as plt
# Data to graph
x = [1, 2, 3, 4, 5]
y = [2, 3, 5, 7, 11]

# Create a line chart
plt.plot(x, y)
plt.xlabel('X axis')
plt.ylabel('Y axis')
plt.title('Simple Line Plot')
plt.show()
```

## 4. Seaborn

Seaborn is a data visualization library based on Matplotlib. Provides a high-level interface for creating attractive and informative statistical charts.

Example:

```
import seaborn as sns
# Example data
tips = sns.load_dataset('tips')
# Create a scatter plot
sns.scatterplot(x='total_bill', y='tip', data=tips)
plt.title('Scatter Plot of Total Bill vs Tip')
plt.show()
```

## 5. Scikit-Learn

Scikit-Learn is a library for machine learning in Python. Provides simple and efficient tools for data analysis and data mining, including classification, regression, clustering and dimensionality reduction models.

Example:

```
from sklearn.datasets import load_iris
from sklearn.model_selection import train_test_split
from sklearn.ensemble import RandomForestClassifier
from sklearn.metrics import accuracy_score

# Load dataset
iris = load_iris()
X = iris.data
y = iris.target

# Split the dataset into training and testing
```

```python
X_train, X_test, y_train, y_test = train_test_split(X, y, test_size=0.3, random_state=42)

# Train a model
model = RandomForestClassifier()
model.fit(X_train, y_train)

# Make predictions
y_pred = model.predict(X_test)

# Evaluate the model
accuracy = accuracy_score(y_test, y_pred)
print("Accuracy:", accuracy)
```

## 6. TensorFlow and Keras

TensorFlow is an open source library for machine learning. Keras is a high-level API for building and training deep learning models, running on top of TensorFlow.

Example:

```python
import tensorflow as tf
from tensorflow.keras.models import Sequential
from tensorflow.keras.layers import Dense

# Create a sequential model
model = Sequential([
    Dense(32, activation='relu', input_shape=(4,)),
    Dense(32, activation='relu'),
    Dense(3, activation='softmax')
])

# Compile the model
model.compile(optimizer='adam', loss='sparse_categorical_crossentropy', metrics=['accuracy'])

# Train the model with example data
model.fit(X_train, y_train, epochs=10, batch_size=32)
```

## 7. State models

Statsmodels is a library for estimating statistical models, performing statistical tests, and exploring data.

Example:

```
import statsmodels.api as sm

# Example data
X = sm.add_constant(X)  # Add a constant for the intercept
model = sm.OLS(y, X).fit()  # Fit a linear regression model
predictions = model.predict(X)  # Make predictions

print(model.summary())  # Model summary
```

These modules provide a solid foundation for most data analysis and machine learning tasks in Python. Each has extensive documentation and a large user community, making it easy to find additional resources and examples.

# Thanks

This book would not have been possible without the support and collaboration of many people. I want to express my deepest gratitude to everyone who contributed to this project:

- **To my family**, for their unwavering support and patience during the long hours of work and writing.
- **To my friends and colleagues**, who provided me with valuable feedback and moral support.
- **To the Machine Learning community**, for sharing their knowledge and contributing to the growth of the field through open and collaborative resources.
- **To developers of open source tools and libraries**, whose dedication and effort have made the implementation of Machine Learning algorithms accessible to everyone.

Without the support and contribution of all of you, this book would not have been possible. Thank you!

# About the Author

Denis Sanchez Leyva is a passionate Full Stack developer and writer with vast experience in the technology field. Born in Cuba and currently residing in Miami, FL, USA, Denis has dedicated his career to exploring and teaching the wonders of web development and Machine Learning.

## Biography

Denis began his journey in the world of technology at a young age, inspired by technology's ability to solve problems and transform lives. With a background in electronics and programming, he has worked on a variety of projects ranging from developing web applications to implementing Machine Learning models.

## Career and Achievements

- **Full Stack Developer:** With experience in multiple technologies, Denis has developed robust and efficient applications that address real problems.
- **Researcher and Author:** He has participated in several research projects and has contributed to the community with articles and books on web development and Machine Learning.

## Teaching Philosophy

Denis firmly believes in the importance of practical and accessible education. Following Ernest Hemingway's theory of being precise and direct, their goal is to make learning simple, easy to understand and applicable to real-world situations.

## Publications

Denis is the author of several books and articles addressing different aspects of web development and Machine Learning, with a focus on making the technology accessible to everyone.

## Contact

For more information, questions or collaborations, you can contact Denis through his social networks or his website:

- **Amazon:** https://www.amazon.com/stores/DENIS-SANCHEZ-LEYVA/author/B0CYXS4R3K?ref=ap_rdr&isDramIntegrated=true&shoppingPortalEnabled=true
- **LinkedIn:** https://www.linkedin.com/in/denis-sanchez-543064167/
- **Youtube:** @denissanchezleyva8954

**June / 2024**

www.ingramcontent.com/pod-product-compliance
Lightning Source LLC
Chambersburg PA
CBHW062312220526
45479CB00004B/1142